GUNFLINT BURNING

GUNFLINT BURNING

FIRE IN THE
BOUNDARY WATERS

CARY J. GRIFFITH

University of Minnesota Press
Minneapolis • London

The University of Minnesota Press gratefully acknowledges the generous assistance provided for the publication of this book by the Hamilton P. Traub University Press Fund.

Frontispiece: Firefighters creating a backburn along the Gunflint Trail. Photograph by Dan Baumann.

Map of the Boundary Waters Canoe Area Wilderness (BWCAW) and Ham Lake fire by Matthew Millett

Published by the University of Minnesota Press
111 Third Avenue South, Suite 290
Minneapolis, MN 55401-2520
http://www.upress.umn.edu

ISBN 978-1-5179-0341-1 (hc)
ISBN 978-1-5179-0342-8 (pb)

A Cataloging-in-Publication record for this book is available from the Library of Congress.

Printed in the United States of America on acid-free paper

The University of Minnesota is an equal-opportunity educator and employer.

23 22 21 20 19 18   10 9 8 7 6 5 4 3 2 1

For Marla Jo Griffith (1956–2017),
who loved the forest and everything in it

and for Ryder, Cormac, and Mylo,
the next generation

The virgin forests of the Boundary Waters region were literally "born in fire." . . . The landscape–vegetation mosaic is like a giant kaleidoscope, with fire being the principal force that periodically rearranges the patterns of vegetation types. . . . In this sense, then, virtually the whole terrestrial ecosystem is fire dependent.

—MIRON "BUD" HEINSELMAN,
THE BOUNDARY WATERS WILDERNESS ECOSYSTEM

The blaze, which became known as the Ham Lake fire, raged for two weeks and crossed the Canadian border. More than 140 structures in Minnesota were destroyed, and 75,000 acres of woodlands burned. Reports estimated damages at more than $100 million. Firefighting costs alone totaled $11 million.

—JANE VARNER MALHOTRA,
DULUTH NEWS TRIBUNE (JANUARY 15, 2012)

CONTENTS

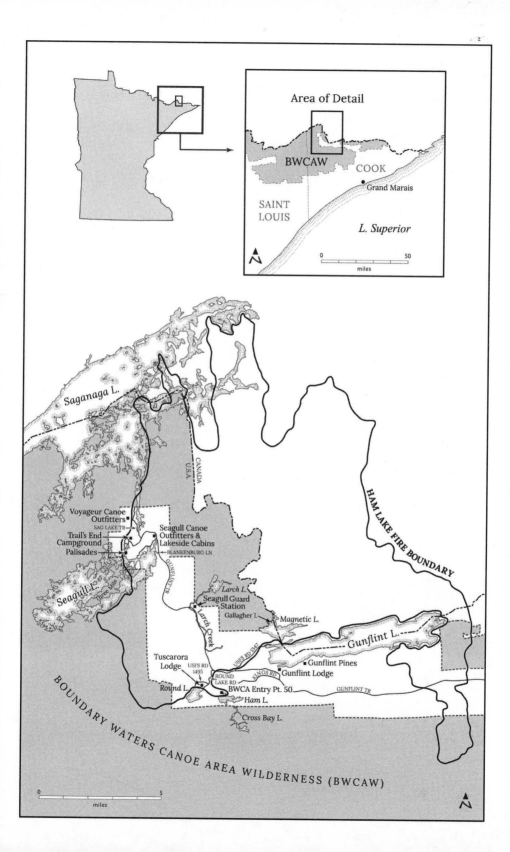

ABBREVIATIONS

BIA   Bureau of Indian Affairs

BWCAW   Boundary Waters Canoe Area Wilderness

DNR   Minnesota Department of Natural Resources

GTVFD   Gunflint Trail Volunteer Fire Department

IAP   incident action plan

IC   incident commander

ICP   incident command post

MIFC   Minnesota Interagency Fire Center

MNICS   Minnesota Incident Command System

NPS   National Park Service

USFS   U.S. Forest Service

PROLOGUE BOB MONEHAN'S PLACE

Seventy-six-year-old Bob Monehan heard a car pull into his drive. He did not need to look to know it was Deputy Tim Weitz. The cruiser's door opened and shut. A moment later Bob heard the deputy's footsteps pounding up his outside stairs.

The two men had been friends for most of the twenty-one years Bob had been living up Sag Lake Trail. Sometimes when Weitz found himself at the end of the Gunflint Trail on a late-night call and he was too tired to drive one and a half hours home to Grand Marais, he would turn into Bob's drive, let himself in, and sleep on a cot in the storage area beneath Bob's cabin.

But both men knew today's visit was no social call.

A distant wall of smoke obliterated the southern sky. Overhead, a Canadair CL-415 SuperScooper, its belly full of water, roared toward the front lines. At the base of that wall a growing array of men, women, and machines were fighting the raging flames. No one expected the fire's march to abate. Bob Monehan's place was dead center in its path.

The deputy entered Bob's house, nodding to his old friend. "It's time."

Deputy Weitz was forty-five years old and in excellent shape, considering he spent most of his time sitting in a cruiser, patrolling the Gunflint. At around five feet eight, with broad shoulders and an athletic build, he looked like the descendant of one of the region's first European visitors, the French voyageurs.

Bob was the same height as his friend. While he didn't have the deputy's shoulder width, he still had plenty of strength and energy for a man almost three decades older.

"Figured," Bob said.

"The sheriff called it. Everyone up the peninsula's gotta get out."

The peninsula was home to more than twenty cabins and the Voyageur Canoe Outfitters. Bordered on three sides by narrow Saganaga Lake waterways, the only way in or out by car was Sag Lake Trail and its connection to the Gunflint.

Deputy Weitz expected Bob to spring into action—hurry into the corners of his home and gather whatever it is people in front of forest fires want to save from the flames. But Bob didn't move.

"Not going," he said.

"Have you looked south? That fire's already knocked out the phones and electricity. The Forest Service says it's headed straight up the peninsula."

"I'll be OK."

"I don't think so. It's a mandatory evacuation."

Bob looked at his old friend. Both of them knew the sheriff—no one, for that matter—could force the old retiree to evacuate. "Not leaving."

The deputy was afraid it might come to this. "You want to burn here with your cabin?"

"We'll be OK. I've got my gear. And the sprinklers have been running for twenty-four hours."

Working off a neighbor's design, Bob built a twenty-plus-head sprinkler system that surrounded three sides of his house, most of his trees, and his outbuilding. Fortuitously, he had set it up and tested it Saturday morning, almost the same time a camper on Ham Lake struck the match that started this blaze.

From what Deputy Weitz had seen and felt from this fire, the untested sprinklers sounded laughable, providing you ignored the gravity of the situation. "I've seen this fire. With this wind and the dry forest, it's burning out of control. That gear isn't going to help, and nobody knows for sure if those sprinklers work. You should get in your car and follow me out like everyone else."

"Maybe . . . but I'm staying."

Deputy Weitz had been to the end of the peninsula and started working his way down. The two cabins north were seasonal lake homes, both empty this early in the season, certain to succumb to the flames. The deputy had others down the road he needed to clear out. Most of the cabins along Sag Lake Trail were south of here, few of them protected by operating sprinkler systems. And these systems had never faced the onslaught of a wind-driven wildfire.

"The sheriff said if you're crazy enough to stay, we need to notify your next of kin."

Bob had four children, three still living; two in Illinois and one in Wisconsin. Weitz had visited Bob's house in Illinois and met them all.

"You know where to find them," Bob said.

The deputy stared at him, then said, "Is your boat gassed up and ready, in case you need to make a run for it?"

Bob nodded. "Yeah," he said.

The deputy let out an anxious sigh, reached for his friend's hand, shook it, and said goodbye. Then he let himself out.

Well to the south, the wall of smoke was huge and impenetrable. Another plane roared low overhead, flying toward the flames. The deputy descended the stairs, dodged through the falling water, and returned to his cruiser. Somewhere inside himself Deputy Weitz said goodbye to his friend, because he did not think Bob Monehan's house would make it through the night. And he did not think his friend was the kind of man who would leave it.

• • •

Bob was born in 1931. A toddler of the Depression, privation was bred in his bones. At an early age Bob learned how to build and fix things. As a young man he fought in Korea and saw things no human should ever see. When he returned, he landed a job as a telephone repairman for an Illinois phone company, where he spent the last twelve years of his career as a supervisor.

The day Bob retired, he headed north. It had long been his dream to live off Sag Lake Trail. Once his acre-plus was paid off, he designed and built his home. It was a simple house; the small, two-bedroom structure had been built into a saddle of rock and sand between two granite rises. On the north side of Bob's house a lake-stone fireplace was set into the wall between two bedroom doors.

His partner, Charlene, had joined him at the house, and until she passed, not long ago, they had enjoyed the place together. Bob had a lot of memories with Charlene, and though he missed her, he knew he was still strong and fit enough to make more.

Bob still hauled, stacked, and burned wood to heat his cabin. The south half of the house contained a kitchen that ran along the back of the cabin, facing west. A dining table was set off the kitchen. The front of the house faced east, overlooking a narrow southern arm of Saganaga Lake. On the other side of the front deck were bird feeders he kept well provisioned throughout the year. The deck was twenty feet from his dock. Moored to his dock was his boat. Across a hundred feet of water, tree-covered granite rose to a large outcrop known as Blueberry Hill. From the promontory of Blueberry Hill, you could see the peninsula and everything around it all the way into Canada. The entire area was surrounded by tinder-dry forest.

Bob was proud the cabin had cost him little more than the price of supplies to build it. Over the years he had done favors for old friends—plumbers, electricians, masons, carpenters, and other workmen like himself. During those years he built up his bank of favors. When he retired and it came time to build his cabin, he cashed in his IOUs, orchestrating friends and fellow craftsmen like a general contractor. And his friends were only too happy to oblige, considering it was Bob and the cabin was in the north woods on a beautiful lake in the center of the wellspring of the North American continent.

When Bob recalled how he had built his cabin with the help of friends he had known over his lifetime, he knew it was much more than wood surrounding a lake-stone hearth built on one acre of land. It was Bob Monehan's place, and he loved it the way he loved his family and friends.

By dusk the air was thick with smoke. Burned forest ash was swirling out of the sky like snow. Bob walked around to all twenty-plus heads. Only the south side of his house was without a steady spray of water and exposed. There was a ten-foot reach of gravel between the house and his fish shed. Atop the shed a sprinkler doused the nearby trees. Other sprinklers, spread closer to the edge of trees twenty feet south, were soaking everything within the circumference of their arcing spray.

When the flames came, Bob decided he would make his stand on the south side, spraying his southern wall by hand. He ran hose to that side of the house, fastened it to an outside spigot, tested it, and then went inside.

From his basement he had retrieved his old, yellow volunteer firefighting gear—the fire-resistant boots, coveralls, and coat. The coveralls and coat were more than twenty years old and insulated with asbestos. They were heavy as a chain-mail suit. He fetched his old hard hat and hauled it all upstairs.

Because the fire had already downed his electricity and phone, everything in his house was operating on one of Bob's generators. His sprinkler system used two tanks, one propane and one gas. When he flipped off his house lights, he peered into the night and saw the red glow to the south, over the tree line at the edge of his woods. Tonight it was much closer and brighter, its menacing maw spreading across the southern sky. Unusually, the night wind buffeted his south-facing wall. Tonight the fire was coming, and the wind would make it burn like bellows on a blacksmith's forge.

• • •

Bob stepped into his firefighting coveralls and pulled on his boots. He put on his coat. The seventy-six-year-old had been running all day, busy with preparations for one more fight. It felt good to lie back and rest. Weariness settled into his limbs, but he knew he must be ready. As he closed his eyes, he reminded himself to stay alert and alive.

One hundred yards south of Bob's place stood his neighbor's cabin. To the side of the cabin there was a 55-gallon gas tank. Bob, exhausted from the day's activities, lay asleep on his couch. While he slept, the steady march of the fire's front line continued to consume everything in its path. When it came to his neighbor's cabin and gas tank, the flames fed on the dry growth and the cabin's clapboard sides. The fire surrounded the tank and began heating the metal casing to a combustible degree, and then an explosion that shook the air all the way to Bob's cabin sent a fireball into the dark sky.

The sound wakened Bob like a face slap. He sat upright, wondering if his own propane tank had blown. Outside it sounded like a locomotive was roaring toward his walls. His windows were filled with an intense yellow light. He looked at his watch and saw 1:30 a.m. and pulled himself up off the couch. He hurried to the door and peered through the glass panes. The sky was filled with smoke and flames. The place where his neighbor's 55-gallon gas tank had been, 100 yards south, was burning like a devil's torch. Across Sag Lake channel it looked like the whole world was on fire. Judging from the spike in flames, he knew the inferno had claimed his neighbor's cabin. Now he could hear the front edge of the yellow rage, emboldened by the taste of fuel, preternatural in the heart of the night.

Bob crossed the room, grabbed his hard hat, and pushed open his door. He was struck by the noise and wind, like a screaming freight train in his face. The fire was close and roaring straight toward him.

He hurried to the side of his house, picked up his hose, peering over his shoulder as he flipped on the water and waited for the spray to gush out. He doused the entire length of his house before turning to have a better look.

The fire had already reached the forested ridge line fifty feet south. The scream obliterated all other sound. The firestorm was creating its own weather system, and now the hot wind drove the flames straight at him, charging through the trees and underbrush. The light and heat was intense, but the omnipresent sprinkler heads continued shooting a phalanx of spray that kept the worst of the smoke and heat at bay, enabling Bob to breathe. He watched red embers rising out of the fierce wall of flames, but the wet air doused them before they could land. The steady thrumming of his sprinkler system was holding, though he could not hear it over the roar.

He turned back to the house, soaking it in a thick spray of lake water, traversing its entire two-story length with his back tensed against the oncoming fire.

When he finally turned, what he saw staggered him. A forty-foot wall of flame silhouetted the trees in yellow light, their branches beyond his sprinkler's spray succumbing to the hungry fire. Twenty-foot trees were lighting the night like giant roman candles. A hot wind fanned the painful-to-the-ears scream.

The heat, light, and sound would have driven most people to drop everything, fight their way to the waiting boat, toss the lines, roar the engine to life, and race up Saganaga's arm into the hoped-for safety of the big open water.

But Bob Monehan wasn't going to leave his place. He turned his back to the fire, ignoring the rising scream, and kept the water pulsing along the south side of his house, hoping he would not need a miracle to survive.

BEFORE THE STORM

1 STEPHEN POSNIAK

No one can say with certainty what Steve Posniak thought or how he felt about his annual camping trip to northern Minnesota. But over the years and on this particular trip, he left a clear trail of where he went, what he did, and when he did it. Also, given the impressions he made on those who knew him, we can make some reasonable conjectures about his travels, actions, and perspective.

On Thursday, May 3, Steve awakened after dawn in the Tuscarora Lodge bunkhouse, in no hurry to roll out of bed into the forty degree cold. There were several empty bunks, a table with some chairs, and his gear spread over everything, waiting for his final packing. This early in the season he had the place to himself, which was one of the reasons he liked arriving in the north woods more than a week before Minnesota's fishing opener. In ten days the area would be filled with hopeful fisherman. But this early in the spring Steve had the place to himself. He might have to suffer through the cold—some years it had been in the twenties—but he seldom saw anyone on his journeys into the Boundary Waters Canoe Area Wilderness (BWCAW) south of Tuscarora.

At sixty-three, he was recently retired from the U.S. Equal Employment Opportunity Commission, where he worked on data security. Like many of the recently retired, he was probably feeling his way into this most recent chapter of his life. Retirees are sometimes uncertain about how their remaining days will unfold. Steve may have worried about having enough money to sustain him and his wife, Jane, through their later years. Regardless, one thing he likely appreciated about this stage of his life was that he was following no one's clock but his own.

It was his twenty-seventh year in the Boundary Waters. The journey was one he had taken almost every spring since he could remember.

After graduating from Dartmouth in 1966, Steve began a lifelong relationship with Minnesota. Following his graduation, the Vietnam War was beginning to ramp up, which was perhaps one of the reasons Steve applied for and entered graduate school in political science at the University of Minnesota. Steve had always enjoyed politics and had during parts of his adult life been politically active. Perhaps it was at Dartmouth or maybe earlier that he cultivated his love of reading newspapers. Among friends and family he was known often to carry a newspaper or have some folded-up section tucked away into one or more of his pockets.

As for many young people in 1966, attending college qualified Steve for a military draft deferment of 2-S. Commonly known as an educational deferment, it gave Steve a temporary delay from participating in the draft, which would keep him from toting an M16 across the rice paddies of Southeast Asia.

After graduating from the University of Minnesota, Steve took a job teaching political science at Carthage College in Kenosha, Wisconsin, where he met his wife, Jane Comings, who taught French at the school.

When he was a boy, he fed his love of being outside with a bike trip from his home in Washington, D.C., to Harper's Ferry, West Virginia, camping solo along the way. Though he never became an Eagle Scout, during high school he was involved enough with Scouting to make the cross-country trip to Philmont Scout Ranch in the mountains of northern New Mexico. And almost every year for the past three decades he had flown from his home in Washington, D.C., to the Twin Cities, rented a car at the airport, and spent the rest of the day driving to Grand Marais and the Gunflint Trail.

On Tuesday, May 1, when Steve flew into the Twin Cities and rented an SUV, starting his drive north, the temperatures across the state were unseasonably warm. In Minneapolis the mercury climbed to seventy-three degrees. In Grand Marais it reached into the low sixties. The weather forecast up north called for a cool Wednesday, followed by another unseasonable spike in temperatures and continued dry weather. In fact, the area was experiencing drought conditions, though the U.S. Forest Service (USFS) had not yet implemented a wilderness fire ban.

After forty-seven miles on the Gunflint Trail, he turned onto Round Lake Road. It was another mile to Tuscarora Lodge and Canoe Outfitters. Sometime between seven and eight that evening he pulled into the parking lot in front of the Tuscarora Trading Post store, where Andy Ahrendt, co-owner of Tuscarora with his wife, Sue, greeted one of their first wilderness adventurers of the season. He let Steve into one of Tuscarora's bunkhouses. In the bunkhouse Steve found two large canvas packs, one green and one brown, containing the equipment and food items he had preordered using Tuscarora's equipment and menu lists. Andy had packed them earlier and made sure they were ready.

On Wednesday, May 2, Sue greeted Steve at 7:30 with a bunkhouse breakfast. Originally, Steve had planned on heading into the BWCAW on Wednesday. Now he decided to postpone it a day, during which he would go hiking and review his gear packs and make sure he had everything he needed. He would also review where he was going and for how long. He let Sue know he would be staying in the bunkhouse another night.

As he started in on his breakfast, he may have sensed it had grown colder overnight, reminding him you just never knew about

Minnesota. In twenty-four hours there was a twenty-degree shift in temperature, typical for Minnesota this time of year. Wednesday never climbed out of the forties.

On Thursday morning Steve finally stepped into the cold room and shook out his bag. Outside there was a ghost of breath stirring out of the south, which would start a warming trend throughout the day. After his brief breakfast, he busied himself with his supplies, which were spread out across the bunkhouse.

Around ten o'clock, Steve wandered down to the outfitter's office, where he met Andy and told he him wanted a permit to enter the BWCAW at the Cross Bay Lake entry site later that afternoon.

As the raven flew, the campsite he had chosen for this year's visit was only a couple miles to the east-southeast of Tuscarora Lodge. But to get there he had to cross a small stretch of water, unpack and shuttle his gear and canoe over a 66-rod portage with two or three rugged hill climbs, and then paddle across a longer stretch of water to a 30-rod portage. At that point he would drop his canoe into a side bay off Ham Lake. His most difficult portages would be behind him, but he still had farther to paddle and one more haul. Once out into Ham Lake's bigger water, he would have to paddle east to a short 17-rod portage and paddle a long waterway that would finally take him to his ultimate destination, the campsite at the northwest end of Cross Bay Lake.

Andy called the camping reservation line and secured a permit for Steve to enter the BWCAW via entry point 50, Cross Bay Lake, later that afternoon. Steve indicated he would be spending Thursday and Friday nights at Cross Bay Lake and then exiting on Saturday, May 5, from his entry site.

Andy and Sue Ahrendt had owned Tuscarora since 2004, when they acquired the business from Kerry Leeds. Leeds had owned Tuscarora Lodge from 1974 to 2004. Leeds first met Steve in 1979, when Steve first used Tuscarora as his disembarkation point into the BWCAW.

From 1979 to 2004, Leeds got to know his annual spring visitor, re-marking both on Steve's intelligence, but also on how he sometimes lacked common sense and good organizational and planning skills, at least when it came to the practical aspects of entering wilderness. He also remembered that Steve had tipped over his canoe in the past, once needing to be rescued by the owners of Gunflint Lodge. Andy and Sue had known Steve since 2004.

Regardless, back at the outfitter's office Andy now reviewed the rules, regulations, and other information on the overnight permit with Steve. There was a list of fourteen questions he read to Steve, to ensure he remembered the rules. It wasn't the first time Steve had read or heard the questions, including the rules about having fires and burning trash, and he indicated, as Andy read them, he knew each of them. Then he signed and dated the permit, perhaps not noticing the personal note Andy had written on the bottom of the permit: "Be careful with campfires."

After Steve finished at the outfitter's office, he returned to the bunkhouse and began repacking the contents of both canvas packs. Then at around 2:30 that afternoon, Andy ferried Steve and his ca-noe and supplies down the quarter-mile gravel road to the nearby BWCAW Cross Bay Lake entry point. The bay, part of the Cross River, was actually the size of a large pond, surrounded by early spring trees. While the country was very dry, the warm weather had begun to swell the tree buds.

Steve was familiar with the portages, paddling, and campsites around Tuscarora, since in the past he had typically stayed at the Cross Bay Lake site, on Ham Lake, or on Missing Link Lake. The first portage out of the Cross Bay entry site was approximately a fifth of a mile, not simple for a sixty-three-year-old single voyageur carrying a 70-pound aluminum canoe and two packs.

Before arriving in the north woods, Steve had prepared for his journey by carrying 40-pound weights up and down the alley rise behind his home. He explained to his wife, Jane, that he needed to make sure he would be strong enough to hoist his canoe overhead and carry it up and down those steep portage hills, eventually lowering the 70-pound aluminum craft to water on the other side. Steve had traversed these first portages often enough to be familiar with the struggle required to cross them. As he paddled away from the dock, he hoped his modest training efforts would pay off.

On some years, many of the larger lakes in the region were still partially ice covered. In fact, only the day before the ice had finally melted off Seagull Lake, as the crow flies just a few miles up the Gunflint Trail. But Seagull was a big lake with more than a hundred islands and large stretches of open water. Lakes the size of Seagull took longer to freeze in the fall, and longer to thaw in the spring.

Here, where the lakes Steve would be traveling were smaller, the ice had been out for at least a week. But the early May weather hadn't warmed enough to raise the water temperature to more than forty degrees. He had to be careful paddling through water this cold. If he capsized and couldn't get to shore within minutes, hypothermia would turn his muscles waxen, and the frigid waters would suck him to the bottom like a rock. So he probably hugged the shoreline, where it would be easier getting to land if the unimaginable happened.

At just shy of six feet and weighing more than 200 pounds, Steve carried extra weight. He had run cross-country in high school and throughout part of his Dartmouth career. But that was during his youth, long ago, and a relatively sedentary work life had added heft to his frame. Just five years earlier he was diagnosed with diabetes. At this point he was still controlling the illness with oral medication, but he wasn't always good about his diet, exercise, or taking his meds.

A stretch of sand shallows at the start of his next paddle required him to get his feet wet. He crossed the narrow lake and shuttled his supplies and canoe over the next shorter portage with no more incident than a renewed dousing of his feet. A little after midday he approached the campsite near the entrance to the large, open water of Ham Lake. The water here was only twenty-five feet across. To the left there was a convenient narrow, sandy beach between two steep boulder rises.

If he was going to the Cross Bay Lake site, this was the halfway point. It had taken more effort and time than he had expected to get to this familiar camping spot on Ham Lake. The prospect of a long traverse of Ham Lake and then hauling his canoe and packs across another portage, followed by even more paddling, must have been unappealing. And he still needed to pitch his tent and set up camp. The idea of relaxing beside a cookstove, or maybe even a fire, and having a warm meal was presumably enough to make him reconsider his plans. With a couple of strong thrusts, he pushed the canoe's bow onto the sand and carefully stepped out.

It was a perfect campsite. Because it was a spit of land, the other side of the campsite made a gradual slope toward the wide Ham Lake water. From this site Steve could sit in the sun on top of the huge boulder bordering the narrow waterway, or gaze out through the trees onto the open lake.

There was a line of black spruce and smaller deciduous trees along the southern shore of his campsite. He hauled his gear out of the canoe, up the small rise, and then down its back side to an open, flat space on the leeward side of the trees. He pulled his gear out of the canvas packs and spent the rest of the late afternoon pitching his tent and setting up camp.

In the center of the campsite stood two strong black spruce, surrounded by the usual assortment of small bushes, saplings, and grass. Away from his tent, up near the center of a slab of granite, the USFS had installed a wilderness fire pit, complete with an iron grate and squared-off sitting logs around three sides of the grill. As the late afternoon faded into early evening, cool air settled over his camp. The fire pit was the perfect place to kindle a flame against the waning day's chill.

This early in a dry spring, campsites like these have plenty of dry tinder sitting around. Presumably, Steve walked around the site, gathering fuel for a fire. Crossing the portages, hauling gear, and setting up camp would make most people feel stiff and tired. Even if touring around the center island of grass and trees and climbing to the big granite outcrop warmed him, he must have been looking forward to a fire and some dinner.

On this evening there was very little breeze. Assuming he had brought a newspaper with him, he would likely have torn off a section of ads and crunched it into a loose ball, tucking it under the grate and placing some of his tinder on top of it. Technically, burning refuse in the BWCAW was illegal. But it was a small section of newspaper, and if he did not burn it, he would have to haul it out, and besides, it was such a common practice—using paper debris to start a fire—the notion it was illegal was like driving over the speed limit. Everyone did it, and as long as you didn't exceed the limit by more than ten miles per hour, you were never pulled over by a patrolman.

Presumably he turned toward the ball of paper, pulled out his matches, and carefully took one out. He leaned down and struck the match. He touched the flame to the paper's edge, and the small fire spread easily across the print, then onto the kindling. Dry pine needles are such excellent tinder they sometimes seem as though they have been dipped in kerosene. The fire made satisfying crackles and pops in the early evening, accompanied with the rising smell of burning pitch. Steve fed the flames with more kindling and watched them catch and burn.

It was a good start. The paper, twigs, and needles burned steadily. He placed his hands in front of the small blaze. The heat was satisfying. He added larger sticks, and the yellow flicker grew.

As the fire burned, he set about using his cookstove to make dinner. Dinner beside a fire in the woods after a journey is one of life's true pleasures. One of the reasons he had come here was to enjoy watching the sunset and the woods darken. It didn't look like rain, but just to be safe he probably took his matches into his tent with him.

When Steve was finally tucked into his bag, he likely took off his glasses and placed them within a hand's reach. It was important to have essential items nearby, where he could easily find them in the morning. He might have felt around in the dark until he touched the matches and set them near his glasses. Eventually he rolled over, pulled the down bag to his chin, and fell asleep.

2. THE WEATHER BEGINS TO TURN

For writer Gus Axelson, Friday morning was the start of a great day. He had recently landed an assignment from *Backpacker* magazine to produce an article about the effects of climate change on Minnesota's boreal forest. Finally, after recruiting University of Minnesota forestry professor Lee Frelich and outdoors photographer Layne Kennedy, he was heading north. Gus, Lee, and Layne were taking a whirlwind tour of the BWCAW. They planned to paddle the remote Seagull–Alpine–Red Rock–Saganaga Lakes loop in just three days.

Lee, an expert in forest ecology whose primary research involved the effects of large-scale blowdowns and fire on the BWCAW's forests, maintained 750 study plots in the 1.1-million-acre wilderness. Lee knew the trees of the Boundary Waters. From any granite promontory (of which there were a legion across the rugged Canadian Shield), Lee could peer over the undulating land and with GPS accuracy identify a four-hundred-year historical record of burns and blowdowns.

Gus and Lee had worked together before. They met when they both were board members for the Friends of the Boundary Waters Wilderness. Lee became one of Gus's primary resources when he needed a better understanding of forest and fire ecology. Layne had also worked with Gus in the past. Layne was a well-known wilderness photojournalist. He had taken photographs in wild places around the world, including photos of Amazonian pink dolphins and on travels by dogsled with Inuit hunters in the polar reaches of Greenland. Like Gus's, Layne's work had been featured in some of the most renowned wildlife publications on the planet. Layne was looking forward to the three-day paddle through spring wilderness, which he suspected would be rugged but easier than his dogsled forays across Greenland's snow and ice or his trip by boat and dugout through the swollen waterways of the largest rain forest in the world.

During the three-day paddle Lee was going to give Gus and Layne a guided tour that would take them through prehistoric and modern catastrophes. He would also introduce them to what he suspected was the oldest tree in Minnesota: a thousand-year-old craggy cedar that could be seen still hugging the northwest shore of Seagull Lake's Three Mile Island, provided you knew where to look.

By the time Gus saddled up his Subaru Legacy with his canoe and camping gear, it was almost seven o'clock. The morning was clear and already warming. It had been a pleasant few days, and he was hoping the weather would hold. Forecasters predicted unusually warm weather all over the state, which was perhaps why—as Gus fastened the green canoe to the top of his car—the wind was already starting to pick up. It was going to be clear, warm, and blustery.

Across town Layne hoisted his yellow sea kayak onto the top of his Jeep Cherokee and stowed the rest of his gear.

At home Lee was making similar preparations, placing his camping gear into the back seat of his Buick Century.

All three were trying to beat rush-hour traffic. Since they each had different return schedules, they were driving separately. The plan was to meet at Voyageur Canoe Outfitters at one o'clock, where Mike and Sue Prom, the proprietors, were provisioning a Duluth pack with all their meals and cooking gear.

Gus loved assignments from big-name magazines like *Backpacker*, which almost always included a modest expense account.

Backpacker would pick up the tab for their meals, which was incredibly helpful, given the craziness of Gus's work schedule and home life. It was Friday, the end of his fifty-plus-hour work week as a writer/editor with the *Minnesota Conservation Volunteer*. Gus's wife was pregnant with their second child. He looked forward to his upcoming adventure but knew he couldn't be gone too long during a time when his home and work to-do lists were growing. Still, considering the rare prospect of a three-day journey through wilderness with like-minded enthusiasts, when he would mix writing business with travel pleasure, he was happy to be on his way.

From the Twin Cities, the drive to the end of the Gunflint Trail took five and a half hours. They were on a tight schedule. They needed to get their provisions from Mike and make sure they were paddling across Seagull Lake's big water into Alpine Lake by supper time, where Gus planned to set up the first night's camp. Before Alpine, Gus was looking forward to hearing more about Lee's perspective on the prospects of their beloved wilderness and to viewing Minnesota's oldest living tree.

Most days the weather across Minnesota is only mildly affected by the climate across the rest of America and the world. As part of Lee Frelich's research into forest fires and blowdowns, he always paid attention to the impact of weather systems on extreme events. For Lee, the *butterfly effect* was more than a scientific proposition. Part of chaos theory hypothesizes that a small change at one place in a deterministic nonlinear system can result in large differences in a later state. In nonacademic terms the theory meant something as simple as the disturbance from a butterfly's wings over China could be part of the cause, perhaps weeks later, of a hurricane in the Gulf of Mexico. The example was extreme and used only for illustrative purposes. But Lee had seen firsthand how climate all over North America and the world was interconnected and how a storm system in one area could have a profound impact on weather more than 1,000 miles away.

On this day, Lee, Gus, and Layne had no idea that 1,000 miles to the south-southwest a major storm system was beginning to unfold. In Kiowa County, Kansas, it was going to be a hot, humid afternoon. The atmosphere would be ripe for a huge midwestern system that would begin forming in earnest after 5:00 p.m. By the end of Friday, the storm would spawn twenty-five tornadoes across the Midwest, including a force EF5 tornado, 1.7 miles wide with winds in excess of 200 miles per hour that would obliterate the town of Greensburg, Kansas.

And on Friday the massive weather system was just getting started. The real show would begin on Saturday.

By the time Gus, Lee, and Layne met at Voyageur Canoe Outfitters, they were further behind schedule than Gus had hoped. The Proms were waiting with their provisioned Duluth pack. Lee and Gus parked their cars at Voyageur and climbed in with Layne, who drove to the nearby Seagull River boat landing. Lee was dressed in a plaid shirt and tan khakis. Wearing wire-rimmed glasses and with short sandy hair, he looked like a college professor in the woods. Gus and Layne both wore the kind of wilderness gear that would breathe, keep them warm, and wick away sweat and water.

Layne pulled his kayak off the top of his Jeep, while Gus and Lee readied Gus's canoe.

It was a beautiful afternoon, already in the sixties, and the three men stared out at the open water, thankful the channel here was narrow and protected and would remain so until they turned their prows into the larger expanse of Seagull Lake. Overhead, the wind was blowing hard and steady. They knew when they entered Seagull's big water—4,000 acres dotted by more than a hundred islands— they would need to navigate carefully to make their way across the lake to the Alpine portage.

The wind was out of the east-southeast, and when Gus stared at his McKenzie map 6 for Saganaga Lake, he didn't like the spread of open Seagull Lake water. Seagull's wide girth angled four miles from the northeast (the point at which they were going to enter) to the southwest. To reach their 100-rod portage, it was a three-mile paddle across open water, traversing almost perpendicular to the wind and waves. And by nightfall, Gus reminded them, if they wanted to keep to their tight schedule, they should be in Alpine lake.

At the landing on Seagull River, Gus reached down and tested the water. The shock of it startled him.

"Mike said Seagull's ice-out was yesterday. That water's thirty-six degrees, and it feels like thirty-six degrees."

Layne and Lee both tested the water and felt the sudden cold ache of it.

"Wouldn't last more than two minutes in that, even with our vests," Layne said.

They donned and tightened their vests, knowing if they capsized and plunged into that water, the consequences could be tragic. It was going to be a careful paddle.

"We need to stay to the leeward side of those islands, to start," Layne said.

Layne was a seasoned kayak paddler. One man alone in a sea kayak was much more stable than two in a canoe. And Layne had paddled Lake Superior many times, which had waters as frigid as these and sometimes boasted waves big enough to surf.

"Definitely have to be careful," Gus said, worried.

While the two were getting their canoe and kayak loaded up, ready to push off, Lee wandered over to the left of the landing. Gus and Layne could see him examining a broken-down cedar tree. Lee called back and explained the tree had been blown over in a previous windstorm, probably hundreds of years ago, and that it had survived in a horizontal position, arching out over the water, putting up new branches from its trunk.

"I've noticed this one before," Lee said. "I think it's around five hundred years old."

Layne walked over and took some photos of the tree, while Gus joined the three, admiring the tree's longevity and tenacity.

Across the river the tops of the trees were swaying in the wind. Lee hoped it would die down, given their impending paddle, but it didn't look promising.

By a little after one o'clock they were on the water, heading south down the river toward Seagull. Layne clipped through the current, enjoying the paddle after the long car ride. Gus and Lee were more cautious. It was the first time the two had paddled together. Gus had taken many trips in his Bell Northwind and knew it was one of the most stable canoes available. But overhead they could feel the wind. Down the river Gus kept the canoe straight and centered. Lee kept paddling in the bow, occasionally stopping his strokes to point out an ancient cedar or a hundred-year-old pine. They talked about the ages of trees in the Boundary Waters. Gus asked him if the five-hundred-year-old cedar was one of the oldest.

"Not by a long shot. The oldest I've found is that thousand-year-old cedar," Lee said. "We don't know for sure, because it's hollow in the middle and we can't count its rings. But it's old. And it's not alone."

After a mile paddle, they came out into the mouth of the big water. They were heading due south. The wind seemed to be increasing out of the east-southeast. A quarter mile across open, choppy water lay Fishhook and Three Mile Islands. They could see calmer waters up ahead where the islands were blocking the wind's sweep. But they would have to cross high waves and whitecaps to get there.

They pushed into the waves and were immediately surprised by the building force of wind. Layne made a quick passage. Gus and Lee were careful, hurried, and watchful. Gus was a skilled canoeist and had to employ forward strokes, back strokes, J strokes, and more to steer their canoe head on into the waves, eventually coming in behind the leeward side of the nearest island.

On the McKenzie map the smaller island directly west of Fishhook was Footbridge. To the west of Footbridge, a scattering of small, rocky, tree-covered outcrops spread all the way to Three Mile Island.

The paddlers used the islands as windbreaks against the blow, making the traverse to Three Mile more manageable.

Gus was hoping to paddle down the southeastern side of the long island to one of the campsites near its middle. Then they could beach their craft and bushwhack up onto one of the island's central high points. Lee had been there many times and knew from that vantage point you could see the forest spread out for miles in every direction, and he would be able to point out the historical blowdowns and burns.

But the southeastern side of Three Mile was open to the wind, and once the canoeists entered the open water, they were hit broadside by the blow. Ahead they could see bigger waves and whitecaps, and it seemed to the three as though the wind was increasing. Realizing it would be easier to traverse the island going down its northwest side, out of the wind, they backtracked and entered calmer waters.

They paddled close to the shoreline until they reached a large cove near the middle of the island. In the center of the U-shaped cove there was a spit of rock with a designated campsite and a place to beach the canoe. Lee had camped here before with students and researchers. A quarter mile farther down was one of his research plots, near the state's oldest tree. It was a good place to pause. The three men decided to take a break from the dicey paddle and wait for the winds to die down. They pulled their canoe and kayak onto shore and shook the stiffness out of their legs.

In the past, Three Mile Island had suffered from several fires. During the fall of 2002 the island was part of one of the USFS's prescribed burns, its effort to reduce the quantity of blowdown fuel that could feed a larger conflagration. Directly behind the campsite was a small wash with a stand of black spruce that survived the prescribed burn, running up to the top of the island.

"Let's check out the top and give the wind a chance to die down," Gus suggested.

The three men started hiking up the wash, toward what Lee

called the Grand Canyon of Three Mile Island. At this point the island was about a half mile across. But the hike up to the topmost part of the island, at 1,500 feet above sea level, took them down a steep canyon wall, into its bottom, then up the other side to one of the island's highest elevations.

Once atop, Lee's promised view was dramatic. The sky was largely clear, and in the open the wind was strong and seemed to be growing stronger. The hikers were hot after the climb, reminding them that the wind and temperature, which had climbed to near seventy, were incredibly unusual for this time of year.

Lee began pointing out and articulating blowdowns and fires that had happened over the past five centuries, concluding with the most recent notable burn, the Cavity Lake fire from the previous July. From this vantage point the path of the 30,000-acre Cavity Lake burn was etched across the southern forest like a huge, chaotic engraving. The path of the burn spread south, more than four miles away, and then to the west and northwest all the way into Alpine.

Lee had plots in Alpine, where they were planning to camp this evening. He was going to show Gus and Layne the effects of a particularly intense wildfire. In Alpine, looking to the northwest, the three could see the far reaches of the effects of that fire. Lee explained that when Cavity burned, the area was dry and covered with fuel from the 1999 blowdown. And the wind was strong. The combination created one of the hottest, most intense fires Lee had ever witnessed.

"If I didn't know better, and there wasn't the lake and water all around us," Gus said, "I'd guess we were in southern Utah." He commented that the region's exposed rock had a pink hue to it. And there wasn't a lot of vegetation.

"It's an ancient lava flow," Lee said. "Two and a half billion years old. This pink granite is beautiful, but it's nutrient poor." Lee explained that the area was crisscrossed by different ancient lava flows. And the flows had all happened at different times and carried different quantities of nutrients, the effects of which you could see in the resulting growth. Down on the North Shore of Lake Superior, it was volcanic basalt, nutrient rich, making those forests thicker and richer.

Lee also pointed out that up until the past mild, relatively wet hundred years, periodic forest fires were a more frequent natural part of the forest cycle.

"Up until around 1890 this forest was mostly made up of white pine, jack pine, and red pine," Lee said. "But in the last hundred years most of it was logged. Then we entered a relative calm, climatologically. The weather grew wetter and milder, and there were fewer fires. In the past those periodic fires would start a whole new generation of pines. But when there wasn't any fire, and then when the Forest Service began suppressing fire, the remaining stands of white, red, and jack pines grew so large that when the 1999 blowdown happened, they were the most vulnerable." During that event, Lee noted, they lost thirty million trees.

From the promontory Lee pointed out other historic fires and their aftermath, etched in the subsequent forest growth. He noted that in some areas black spruce, poplar, birch, and hazel had come in after the loggers took all the mature pine. It was a different forest than it had been a hundred years earlier. But there were still places you could find the old growth.

GUNFLINT LODGE—According to the USFS Air Resource Management Program mission statement, the purpose of the air resource management team is, in part, to "protect air quality by working with industry and regulators, monitoring air and the resources affected by air pollution, and by providing the public with information about air quality." The team is comprised of "a network of technical and policy-orientated specialists with background in biology, engineering, physics, ecology, botany, hydrology, soil science, and forestry." Coincidentally, on this day, the team was finishing one of its national meetings at the historic Gunflint Lodge, a little less than five miles down the Gunflint Trail from Tuscarora Lodge and Canoe Outfitters. The national meeting concluded around noon.

Understandably, these professionals were interested in wilderness areas. Several of them decided to avail themselves of this opportunity to visit and explore one of the nation's premier wilderness areas. Since one of the specialist's office was in Duluth, his wife also decided to tag along.

The group of four men and four women worked with Bonnie Schudy, the general manager of the Gunflint Northwoods Outfitters group, to arrange the details for the eight travelers to enter the BWCAW. They needed four canoes, gear, food, and the necessary permits to spend a couple of nights in the area. They also needed transportation to their put-in site, entry point 50, Cross Bay Lake, just up the road. They decided to set up their first camp on Ham Lake. From that location they could paddle and portage farther south and explore some of the other lakes in the region. This early in the season they figured they would have the place to themselves and were looking forward to their brief wilderness sojourn.

By two thirty the group of eight had left the Gunflint Lodge, headed to their put-in point. The wind was out of the east-southeast at 10 miles an hour, but sometimes gusted up to 20. The temperature would rise to sixty-six, almost ten degrees warmer than normal for this time of year.

For a wilderness trip, four canoes were practically a flotilla. To lessen the impact of groups on the BWCAW, the USFS limited the size of any group to no more than nine people and four watercraft. The big group took a while to put in at the entry point, paddling across the small stretch of water to their first portage. Even though the small water was relatively well concealed, the wind was blowing across it and more or less directly at them, making the paddle strenuous.

Over the next hour the group covered both portages and paddled into the side bay off Ham Lake. Two women in the group were in the first canoe off the portage, so they were the first to go through the narrows into Ham Lake's bigger open waters. But the wind was blowing so hard they back paddled to calmer waters before making another try.

They were surprised to see someone already occupying one of Ham Lake's campsites and spent a moment observing the gentleman, waiting for the others to catch up.

Steve Posniak was walking around the point. The canoeists saw a man who appeared to be around sixty years old. He had grayish-white hair, wore glasses, and was big and stout, with a kind of beer belly. Surprisingly, he also wore headphones. His aluminum canoe was hauled out of the water and turned over. He had laid his socks and boots on top of one of the big boulders on the site, presumably to dry. They noticed other articles of his, scattered around what they could see of the campsite.

The two air-quality specialists were curious if the other campsites on the lake were open, so one of them called out to Steve.

"Have you seen anyone else on the lake?"

Steve pulled off his headphones and said, "There was a group I saw earlier, but they're way ahead of you, and I don't think they're staying on Ham Lake."

They exchanged a few more pleasantries, and then the two women crossed back into the open water and redoubled their efforts against the wind, paddling a quarter mile down the shoreline, where they beached their canoes at the next campsite.

The other canoes followed them, though no one exchanged any more words with Steve. They, too, made similar observations of Steve and his campsite; there were articles spread around, he was wearing a red shirt, was probably around sixty, was heavyset, wore headphones, and so on.

While the next campsite down had a slope of rock sticking out into the lake, even when on the end of it, near the water, the trees and brush at Steve's site prevented clear visibility of Steve, his tent, or the rest of the site.

THREE MILE ISLAND, EVENING—By the time Gus, Layne, and Lee arrived back at camp, they had begun to realize the inevitable. The campsite was on the leeward side of the island, but because of previous island burns, the wind swept over the top of the rock. The breeze headed out across the big expanse of Seagull Lake, creating such big waves and whitecaps that it would be foolhardy to head back onto the water.

The three unpacked their gear, growing increasingly hungry after their paddle and hike. It was so blowy Layne had to move his tent from a rocky outcrop down into a ravine with some ground where he could anchor the nylon with stakes. Lee and Gus pitched their tent away from the water. Because the wind didn't seem to be calming, Lee searched out four heavy, big, rounded stones they could use to anchor the inside four corners of their tent. While he was looking for suitable stones, he came across a birch sapling.

"There's a sign of global warming right there," he said.

Gus saw a healthy young birch tree. When he looked at the professor with a question, Lee explained that it was budding weeks ahead of schedule. Longer growing seasons were not good for birches because that meant the soil was also getting warmer, and birches can't tolerate warmer soil temperatures. It was just another example of how the warming trend was impacting the boreal forest.

By five o'clock they had pitched their tents.

Once the tents were set, the three men happily opened the Duluth pack, searching for what the Proms had chosen for their first wilderness supper. Expecting to find something freeze-dried, they were ecstatic to pull out three fresh steaks, wrapped in plastic, thawed and ready for the grill. It was going to be a wonderful wilderness feast.

While Gus searched for a fry pan, Layne and Lee tried to start a fire. There was a fire pit on the spit of granite running down into the lake. Every time they managed to kindle a flame the strong wind blew the kindling into the open water. Finally, they gave up and resorted to camp stoves. The stoves took longer, but the wait only enhanced how well the steak tasted in the growing dusk.

"I don't think this wind is going to break," Gus said.

"I think it's gotten worse," Layne agreed.

"It's the weirdest early May I've been up here," Lee said. "I'm surprised, with how dry it is, the Forest Service hasn't implemented a fire ban."

The three men talked for a while about the USFS and why fire bans were problematic. Once implemented, the USFS was slow to lift them. And with the Minnesota fishing opener one week away, they were probably reluctant to ban what many fishermen looked forward to: sitting around a campfire in the north woods. The resort owners appreciated their reluctance to call for a ban, at least for now. And it was still well before the traditional start of the fire season, even if the wind was increasing, and the forecast for tomorrow was for some of the warmest temperatures on record. The three men guessed the USFS thought they had time, given how early it was.

Finally, not long after dark, they turned in and tried to sleep through the night.

Gus worried about their trip. They were already seriously behind schedule, and if they were going to make the entire circuit, they had to paddle like hell. And Lee still needed to show them some of his research plots. Gus felt anxious in the growing dark. As if in concert with his feelings, the wind did not diminish, and all through the night he periodically awoke to the sound of their nylon tent flapping like sails in an approaching storm.

HAM LAKE, EVENING—Back on Ham Lake, on the point stretching out between the lake's open water and the narrow entrance to the bay, Steve Posniak was settling in for the night. Apart from the wind, which throughout the day had stiffened as it blew across the open stretch of water, it had been beautiful on the lake. Steve had tried to paddle earlier in the day, but the wind and waves forced him back to camp, where he holed up reading the paper he pulled out of one of his pockets.

A quarter mile farther down the shoreline, the group he had

seen earlier put in at the next campsite. By around five o'clock he could hear them getting out their gear, pitching tents, and beginning to set up their site.

Steve spent the dusk perusing that paper section again, and now he reminded himself that in the morning, when it was calmer, he could use it as starter fuel for twigs and sticks to build a small fire against the chill. And since tomorrow was his day to paddle out, he would also gather all the rest of his paper and burn it. He was looking forward to the sudden warmth he knew it would provide.

Steve could not hear them, but if he had, he would have known that after dinner the USFS group sat around a fire and talked about how best to spend the next day. Since the wind was still a little strong, they decided it would make more sense to keep their current location as their base camp and make smaller day trips into the outlying wilderness areas.

By nine thirty or ten, everyone had called it a night and turned in.

It was a windy night, but Steve was snug in his tent behind the trees. Sometime well after dark he dropped off to sleep.

VOYAGEUR CANOE OUTFITTERS, END OF THE GUNFLINT—For Mike and Sue Prom, it had been a busy, strange day. Outfitting Gus, Lee, and Layne was just one of the details in a day filled with the usual assortment of tasks involved with getting geared up for the season. Fishing opener was only a week away, though the opener wasn't a big weekend at Voyageur. The Seagull River ran in front of the resort and was a natural spawning site for walleye. This early in the year the iconic fish were starting to come into the river and spawn. Consequently, the DNR did not allow fishing on the river until the Friday before Memorial Day, which was when the real fishing season started to get busy for the Proms. But there was still plenty to do in preparation for the coming summer season.

Mike and Sue had been managing Voyageur since they acquired the business in 1993. If you wanted a remote canoe outfitter, perched on the edge of the vast BWCAW and Canadian Quetico, you couldn't get much farther up the road than Voyageur. From Grand Marais, you drove the Gunflint Trail almost sixty miles to its end. Then you followed the signs up Sag Lake Trail to Voyageur's gravel parking lot. If you boated or paddled a mile in almost any direction, you were in wilderness or Canada or both. Voyageur was ten miles farther up the road from Tuscarora.

Among her many other duties—mother, co-owner, outfitter, and more—Sue Prom blogged. Almost every day she managed to carve out a spare few minutes to make an entry. In fact, for these first few days of May, she had penned fairly long posts.

On May 1 her blog entry described the prescribed burns the USFS was planning for the 2007 season. In part, she wrote,

> The USFS is planning a number of prescribed burns during the 2007 season. Severe drought conditions that exist in some parts of Minnesota do not exist in our area thanks to some late winter snowfalls and early spring rains. If these conditions hold steady then the USFS will attempt a number of prescribed burns in the Gunflint Ranger District this spring.
>
> There are currently five prescribed burns proposed for May and June in the Gunflint District.

Sue listed the five areas the USFS proposed to burn, totaling approximately 4,300 acres. She concluded her post with the USFS's reasoning for the burns:

> These prescribed burns are done in an effort to reduce fuel loads left behind from the blowdown in 1999. The success of the prescribed burn program was proven during last year's Cavity Lake Fire when numerous homes and private property was saved due to previous prescribed burns in the area of Seagull Lake.

Sue often included public service announcements in her posts. Presumably her blog was read not only by others up and down the Trail, but also by anyone who had outfitted with Voyageur or planned to in the future. Her entry on May 1 concluded with a public service post for anyone interested in learning more about the USFS prescribed burns:.

> Prescribed Fire Public meeting on May 5, at Poplar Lake Fire Hall
> at 3:00 p.m.

On Wednesday, May 2, Sue had a very long post about a Quetico Outfitters meeting her husband, Mike, attended in Atikokan, across the border in Ontario. She included several news items in which every outfitter up and down the Trail would be interested.

On Thursday, May 3, she had a brief post about the exploding deer population in the area, concluding with the following:

> The other day Josh [Prom] found a treasure while out on a walk.
> A full deer skull with an 8 point rack and he is quite proud of his
> first 8 pointer.
> The deer are plentiful on the Gunflint Trail so be sure to
> drive with care.

Finally, on Friday, May 4, the same day Gus, Layne, and Lee became stranded on Three Mile Island, she posted a press release about a thirty-five-year-old Greeley, Colorado, man who flipped his canoe while paddling Dowdy Lake. There were three in the canoe; the only one wearing his life jacket was a ten-year-old boy. The boy and his father made it to safety, while the boy's uncle succumbed to the frigid waters. It was a terrible reminder of the tragic turn of events that could occur if you were not prepared for the unexpected.

Those living along the Gunflint Trail are familiar with tragedy. If the Greeley man had been in an urban area with people and emergency services nearby, he might have survived. But presumably he was visiting the area because of a love of the solace and respite a wilderness area provides.

The timeless struggle for those visiting and the local residents living at the end of the Trail was how to retain the pristine beauty of the area while enjoying the benefits of civilization. Refrigeration, electricity, phones, and services like local fire and police were often slow to make their way into or near wilderness areas. In many instances local residents feared such conveniences would permanently alter the character of a place. Regrettably, sometimes it takes a disaster to change or foment public opinion in one direction or another.

While the Gunflint Trail now has many of the referenced services, including a robust volunteer fire department, it was not always so. In the 1980s there had been some talk and suggestions—especially by the USFS—about the region creating and supporting a volunteer fire department. But the prevailing sentiment was that it was either not needed, that it was too difficult, or that residents lacked the knowledge, manpower, and finances to create and support it. Then in the early morning hours of July 12, 1991, a historic tragedy occurred that—at least with regard to public sentiment about creating a volunteer fire department—shifted perspectives.

3 THE BURNING OF WINDIGO LODGE

Sometime during that period of night when most people have entered their third, final, and deepest REM state, a fire started on the first floor of Windigo Lodge. Bruce Wayne Kellerhuis, one of the guests, left the lodge at 3:30 a.m. At that time there was no fire. But less than one hour later, Duane Anderson would awaken to screaming and glass breaking and a blaze out of control.

The source of the 1991 Windigo Lodge fire is unknown. Regardless, once the fire began, the dry, wooden structure caught and burned at a voracious rate, rapidly filling the air with dense, suffocating smoke and heat.

The interior of the three-story structure had a classic rustic northwoods look, with huge white spruce logs big enough to support the second and third floors. A pair of stuffed black bear cubs were poised in their climb up one of the two-foot-thick poles. With a large fireplace, ¾-inch tongue-and-groove aspen paneling, and a wraparound deck that overlooked Poplar Lake, it was difficult to ignore the fact you were on the edge of a wilderness area.

The main floor also had a dining room and rec area, a kitchen, a laundry, and two sets of open stairs. The second floor had ten rooms and two bathrooms and on this night housed four employees and six guests. The top, third floor contained the living quarters for the Ekroot family (the owners), and on July 12 also housed two guests, Glen Ray Wittnam and Thomas Frederick Cooley. Charlette Geraldine Ekroot and her mother, Gladys Lillian Merril, were asleep in their own rooms.

The lodge also had three external cabins and a sauna. Duane Anderson was staying in one of the cabins, and it was he who was awakened in the dark by a family member to the sounds of screaming and glass shattering. The sun was scheduled to rise at 5:15 a.m., so it was still dark when Anderson dressed and ran to the lodge and witnessed people jumping from windows and the lodge already in flames. He ran to the west side of the building and entered the enclosed stairway vestibule, where he knew there was a phone. By this time it was so hot, dark, and smoky he could not find it. He ran from the building to his pickup truck, retrieved a flashlight, and returned to the west exit. With the help of the flashlight he managed to dial the Cook County Sheriff's office.

It was 4:21 a.m.

In 1991 there was no Gunflint Trail Volunteer Fire Department (GTVFD). Twelve miles up the road from Windigo Lodge was the Gunflint Trail Fire and Rescue Squad, but it was only equipped to provide emergency medical services. In fact, the nearest organized firefighting capability was in Maple Hill, approximately thirty miles down the Trail toward Grand Marais. Recognizing the absence of the ability to fight fires, some of the local resort owners had acquired portable pumps with hose, but that was the extent of their firefighting equipment and capabilities.

As soon as the sheriff's office received news of the fire, the dispatcher relayed the information to nearby resort owners.

Meanwhile, the interior of Windigo Lodge was in chaos. Owner Vince Ekroot slept on the first floor, confined to a wheelchair because of multiple sclerosis. Presumably he was fast asleep when the fire started. Ekroot was probably the first to succumb to the smoke and eventually flames.

Because of the open stairways, the fire and smoke quickly rose to the second and third levels.

The lodge had recently undergone (and was still undergoing) some renovation, which may have accounted for why none of the smoke detectors present during the last inspection (1988) were affixed or operational. Some guests recalled seeing one behind the bar on the first floor, but it must not have functioned properly, because no one heard an alarm. There were no fire evacuation plans, so none of the employees had ever practiced an evacuation. And the lodge did not have a sprinkler system.

The *Chicago Tribune* reported on the fire ("Minnesota Lodge Burns: At Least 7 Die, July 12, 1991"), quoting some of the survivors and others who arrived at the fire not long after it started. William Nelson, a twenty-five-year-old carpenter and Windigo Lodge employee, was in his room on the second floor when he awakened to the mayhem:

> "I was going to run out the door, but fire came up the stairs, so I
> ran back. I couldn't breathe anymore, so I hit the deck and started
> crawling. And I seen a light out a window, and I threw a chair
> through the window and jumped out. That's when I was hurt."

By this time other people in the lodge had awakened to the fire and smoke and also tried to find their ways to safety. The smoke was so thick some people tried to crawl out, while others—hell bent on finding a way out—tripped over them. No one could see a clear exit.

Nelson recalled his friend Vince Liestman, also an employee, running back and forth, trying to help people navigate the smoke, heat, and flames. Ultimately, Liestman was overcome and perished in the fire.

Nelson, who managed to get out, suffered burns and lacerations. From outside he watched as others still trapped in the structure tried to leap out windows to safety:

> "There were people jumping from the third floor and the
> second floor," he said. "I saw three people jump, and it was

terror, a feeling of helplessness. You couldn't help them.
You could hear people screaming inside, and I heard people
screaming outside . . ."

After the sheriff's dispatcher alerted resort owners, one of the first
on the scene, arriving just before dawn, was Bruce Kerfoot, owner
of the Gunflint Lodge, approximately thirteen miles down the road.
Kerfoot was also a member of the Gunflint rescue squad.

"It was an inferno," he said. "It was visible for maybe five miles
away. The smoke plume was perhaps 400 feet high, and flames were
in excess of 50 feet."

Another member of the rescue squad, Daniel Baumann, said
that although it was raining when he arrived, the fire was too far
along to be extinguished.

Because the stairwell was open all the way to the third floor and
the smoke and heat rose rapidly, the third floor was overcome almost as quickly as the second. On July 14, 1991, the *Orlando Sentinel*
ran a story ("'Instinct' Made Lodge-Fire Survivor Jump") in which
Glen Wittnam, one of the two survivors from the third floor, said
he was awakened by a voice and the smell of smoke. He awakened
his roommate and then turned to the room's window. He broke the
screen and window frame and jumped.

"I didn't have time to think. . . . It was pure instinct."

He landed on the deck, breaking both feet and an ankle, and was
in stable condition that Saturday at a Duluth, Minnesota, hospital.

His roommate, also from Illinois, did not escape and was pre-
sumed dead, Wittnam said.

"Had I stayed in that building another thirty seconds I would
have been dead," Wittnam said from his hospital bed.

Co-owner Charlette Ekroot also must have arisen to smoke and
fire and managed to open her bedroom window and jump, but not
without suffering bone fractures.

When the ashes finally settled, gone were Vincent Rolland
Ekroot Sr., Gladys Lillian Merril, employee Vincent Charles Liest-

man, and guests Michelle Lynn Swenson, Greg James Swenson, Brian Jay Porter, and Thomas Frederick Cooley. Injured while escaping were co-owner Charlette Geraldine Ekroot, employees Adam Troy Maxwell and William Joseph Nelson, and guests Milan Frank Matetich and Glen Ray Wittnam. In the ensuing chaos, Duane Anderson, who first called in the fire, also was injured. Guest Donald Jay McComb escaped without injury.

Resort owners brought their pumps and hoses, but by the time they arrived, the fire was already so well under way they used two of the pumps to keep the flames contained and prevent them from spreading to nearby structures and forest.

As with most tragedies, there were a string of consequences that lay like stones in a path toward its terrible conclusion. If the fire had not struck at the deepest hour of the night. If the lodge had operational smoke detectors in every room. If its infrastructure was built of something other than heavy timber construction with exposed beams. If there had been a fire department within five miles of the lodge. If there had been a way for guests to easily exit the three-story structure's interior or an external fire escape or an automatic sprinkler system. If fire doors had been installed to block stairways so they didn't operate as open channels by which the smoke and flames could easily and quickly ascend. If employees had understood and practiced fire evacuation plans. If *any* of these things had been true, the outcome of this devastating night might have been something other than seven deaths, six injured while escaping, and only one person leaving the scene without injury.

If tragedies have silver linings, this one's may have been that eventually what happened the night of July 12, 1991, would have a profound impact on the future of structure fires all up and down the Gunflint Trail, especially those that were to be in the path of a blaze that started sixteen years later.

FIRE DAY ONE

4 PREPARATION

Even though Steve Posniak could probably sense it was going to be another unseasonably warm day, one of the first things he may have considered, awakening in his tent, was building a fire against the morning chill. It was almost seven o'clock, and the sun had been up for more than an hour, with the temperature already climbing into the forties. But if you awaken on the ground in a tent in that kind of cold, it is hard not to think about starting a fire to ward off the chill.

The nearby Ely weather station logged a sunrise of 5:43 a.m. CDT with a temperature of forty-four degrees. The average low for this time of year was just above freezing (thirty-three degrees), so this day began warm, a harbinger of the day's rise into the sixties. The other weather phenomenon Steve must have noticed was the wind. Normally dusk to well after dawn marks the cessation of wind, as though the atmosphere needs time to rest. But throughout the evening of May 4 and the early morning of May 5, the atmosphere was restless, blowing at a speed of at least 5 miles per hour. It was enough to occasionally rattle Steve's tent flap, though presumably he managed some fitful rest.

Now, in the early morning, there was still a slight breeze. Not too windy. But by seven o'clock, it was already starting to pick up. For a while he listened to the wind. He was snug and warm in his down bag. The wind flapping his tent fly was cold, this early in the morning. He thought about getting up. He may have thought about reading some of the material he had carried with him.

From farther down the shoreline, the voices of some of the USFS people likely traveled across the lake. He probably heard them, though not what they said. When the wind is in the right direction, sound can travel unabated across the water's surface.

After two nights on Ham Lake, he was scheduled to paddle out, leaving via the two portages and waterways he had traveled to get there. He may have thought about taking another paddle across Ham and maybe into Cross Bay Lake, but if the wind kicked up like it had the day before, paddling anywhere but out the more or less protected waterways back to his entry point might be foolish. The water was extremely cold and could be dangerous, if the unimaginable happened. Besides, he wanted to burn all his paper trash prior to leaving. The earlier he burned it, the better, given the wind. At this point he was probably looking forward to the warmth the flames would provide.

SEAGULL LAKE, EARLY MORNING—All night the blow continued out of the southeast. While Layne, Lee, and Gus were on the opposite side of Three Mile Island, the open escarpment gave little protection against the wind. It was cool, but apart from the wind, they were experiencing the same start to the day as Steve Posniak (twelve miles to the southeast); the temperature was forty-four degrees and starting to climb.

Since the wind was out of the east-southeast, they knew it was going to get warmer. They got out of their tents and tromped around, coming awake in the early morning cold.

Gus was mindful of their lack of progress. "If we're going to finish that loop, we'd better make some serious time today," he said.

"We're going to have to paddle like hell," Layne agreed. He knew he could keep up in the kayak, particularly if there were waves. When he peered out over the big northwest expanse of Seagull's open water, it appeared to him as though it was going to be dicey again, particularly considering that at this hour he could already watch the sweep of wind frazzling the lake's surface to the northeast.

"Let's see what Mike packed us for breakfast. I'm hungry," Gus said. He rummaged through the Duluth pack, searching for their meal.

"Before we go, we still need to check out the oldest tree in Minnesota," Lee said.

"Definitely," Gus said.

Layne agreed. They were only a quarter mile away, and they could do a quick tour of one of Lee's research plots before striking out across the big water and into Alpine. Beside the plot, close to the Seagull Lake edge, rose the thousand-year-old cedar. Lee was looking forward to greeting his old, gnarled friend and checking on the progress of his trees.

The three quickly prepared and ate breakfast. Afterward they began breaking camp. Glancing out over Seagull's big water, they could see that the lake's surface was bumpy but passable, at least for now. The sooner they could get across the big bay to the other side, the better.

USFS GUNFLINT RANGER DISTRICT OFFICE, GRAND MARAIS—Kurt Schierenbeck, district fire management officer for the Gunflint District of the Superior National Forest, was a thirty-plus-year veteran of the USFS, almost all of it spent in *fire*. On this morning he awakened early, thinking about a cup of coffee and his day. He was on duty today and had to head to the USFS Grand Marais office on the southwest side of town. In the early morning, with a mug of coffee in hand, Kurt peered out his front windows and noticed Lake Superior was already choppy from the wind.

Early May is the start of the fire season in northern Minnesota, and this Saturday morning the signs were worrisome. Kurt was serious about his work with the USFS. He was a recognized expert and held the Red Card that indicated he was a Type III Incident Commander. The card told others in the USFS he had a lot of class time, fieldwork, and fire and disaster management experience.

According to the U.S. Center for Excellence in Disaster Management and Humanitarian Assistance, the Incident Command System is "a set of personnel, policies, procedures, facilities, and equipment, integrated into a common organizational structure designed to improve emergency response operations of all types and complexities. ICS is a subcomponent of the National Incident Management System (NIMS), as released by the U.S. Department of Homeland Security in 2004." In instances of natural and human-caused disasters, including wildfires, incident commanders (ICs) are the leaders of personnel and equipment. ICs are rated from Type V (which in a wildfire situation might be an unattended campfire) to a Type I (the most severe and complex wildfire). Kurt was a Type III IC, which requires the oversight of extensive equipment and people to fight a serious wildfire.

In the morning, Lake Superior's waves caused Kurt to do what he often did during fire season. He walked out onto his back deck with his mug of coffee in hand and considered the day. First, he noticed the wind. It was already strong, and that wasn't good for wildfires. Second, he noticed his backyard grass. Normally his morning grass was covered with dew. He bent down and ran his hand over the grass. His fingers came away dry.

Low humidity, another bad sign.

Kurt reminded himself to stop by the DNR office on his way into work, where he would review DNR and USFS firefighting resources and talk about how to respond, should it become necessary. Because it was dry and windy, he worried that if a wildfire occurred, it would be a challenge to control.

May 5 was early in the fire season, but since it was still the season, he didn't get into his normal USFS uniform; he donned his firefighting Nomex gear. The yellow shirt and green pants are relatively lightweight but fire resistant. Anyone fighting wildfires is required to wear them. In the early morning he hoped it wouldn't be necessary, but if it was, he would be ready.

The Schierenbecks lived two miles northeast of Grand Marais

on Highway 61. The DNR office is on the northeast end of town. On his way to the USFS office, he pulled into DNR to have a talk with his firefighting colleagues. On this day Tom Lynch was the DNR's Type III incident commander, if a serious response of that level was needed. The response to wildfires is largely dictated by where a fire begins. If a wildfire begins on DNR land, the DNR usually takes the lead in responding. Northern Minnesota is covered primarily by federal land, which would dictate the USFS take the lead.

On this morning Tom Lynch and fellow DNR colleague and firefighter Gary Jorgeson sat down with Kurt and reviewed the combined equipment and personnel resources that were available. The DNR, like the USFS, had bulldozers, planes, fire engines, cache trucks, and a wide assortment of different kinds of firefighting equipment. The three men ran through their lists of who was on duty. Today the fire duty officer for the DNR was Tom. For the USFS, it was Kurt. Vance Hazelton, a USFS firefighting colleague of Kurt's, was also on call in case anything happened. Vance was a Type IV incident commander. If there was a fire, Vance would be the USFS person to respond on the ground.

At the DNR office, the three men discussed how the signs were all unfavorable. In addition to the wind and low humidity, the temperature was rising. Kurt thought the most logical place for a fire to start this early in the season, when there were still patches of snow in places off the Gunflint Trail, would be down by the lake, where there was plenty of open, dry grass. Tom agreed. They would be on the lookout for something to start down near Superior. The men said goodbye, and Kurt walked out to his truck and headed toward the USFS offices on the southwest end of town.

HAM LAKE, MORNING—Sometime around eight o'clock Steve must have pushed out of his down bag enough to pull on his Dartmouth sweatshirt, and then finished wiggling out, pulling on his cold pants, turning to unzip the tent fly and reaching for his boots. Boots that sit out in that kind of cold can feel like two clumps of ice. Presumably

Steve's hands were finally starting to come alive, but his fingers were still stiff and achy. He managed to pull on his boots and then carefully backed out of the tent on all fours.

The sunlight must have felt good, but it was definitely still chilly. There were some clouds, but the sky and occasional sun were promising another excellent day. Must have felt like it, anyway. The breeze from the east-southeast created a small chop on the lake's surface, but as he got up and tramped around his site, looking for twigs and kindling to hold and continue the flame from a piece of newspaper, he might have noticed the small channel leading into Ham Lake's secluded bay. His campsite's spit of land protected it from the southeasterly breeze, and it was calm.

There was a line of black spruce and smaller deciduous trees along the southern shore of his campsite. He had pitched his tent on their leeward side, and now their branches were ruffling in the late morning, holding back some of the wind. But this early in the season it was a leafless barrier, and Steve must have felt the breeze through the trees and over his campsite, heading to the west-northwest.

If Steve wanted to build a quick fire to give his coffee and breakfast preparations a little warmth, he would need to gather some dry tinder. Touring around the center island of grass and trees, picking up sticks, leaves, or pine needles and climbing to the big granite outcrop would have warmed him. If he wanted to build the flame into something that provided solid warmth, he would have searched for larger branches to catch and hold a flame. The area was so dry it was probably difficult to distinguish between dried, dead wood and what was green. He could have gathered several larger branches, some possibly with green pine needles still bunched along them, and dropped the armload of fuel beside the fire grate.

The breeze must have increased the morning chill. Within the square of log seats, the wind would have been partially obstructed by the line of trees along the lakeshore. Oddly, the obstruction caught the breeze, and if Steve had followed it, he might have felt a gentle eddy in the campsite, circling back and to the right around the center island of trees and brush. But when the breeze was this light and to his back, the only thing Steve probably felt was a cool whisper on the back of his neck. He must have been looking forward to a warm fire.

Because he still had paper to burn, he probably used some of it to build a paper tent surrounded by dry kindling. Presumably he used his body to create a small windbreak, struck the match, touched it to the edge of paper, and watched as the flame blossomed and caught on the twigs and dried pine needles.

A wind-breath flare spread across twigs and sticks would have brought the fire to life. Here, where the ring of rocks created another barrier against the full force of the breeze, the paper, twigs and needles must have burned steadily, warming his hands. If he began feeding in some of the larger branches, watching them catch and burn, he would have seen the fire grow and felt the heat rise up his arms and onto his face. Fed by the light wind, the blaze would have fired and grown quickly.

At some point Steve must have placed larger branches, some possibly with green needles, directly on top of the flames. If so they may have started smoking, heavily, sending up a thick gray cloud above the site. If the smoking needles suddenly reached a combustion point and burst into larger flames, the conflagration could have added more smoke into the rising light breeze. And if it suddenly seemed dangerous, Steve could have doused it with nearby water from a pot, causing even more smoke to rise into the morning air.

Something like this must have happened, because as he stood on top of the boulder over the narrow waterway, two canoeists suddenly appeared, paddling toward the larger lake. They glanced at him briefly but did not stop paddling.

The two paddlers had put in earlier in the morning at the same BWCAW entry point as Steve had two days earlier and the USFS air resource people just yesterday. These canoeists had also worked with Bonnie Schudy from Gunflint Northwoods Outfitters. They had arrived from Ohio the night before, staying at the Gunflint Lodge. They wanted to get an early Saturday start, so Bonnie dropped them off with two packs of gear and the canoe and paddles before eight that morning. She noticed they were efficient. By the time she headed back, it was around eight, and they were already paddling away to their first portage.

By now it was around nine, and they had already hauled their gear and the canoe over the two portages and made the subsequent brief paddles. But they had not yet hit the open wind head on. They could hear it and feel it and knew as soon as they exited the narrow inlet beside Steve Posniak's campsite, they would face it head on.

They noticed the smoke plume and Steve standing on the rock, and clearly Steve saw them. But neither said anything to the other or made a customary nod or wave. The Ohio canoeists were intent on paddling. As they hit the larger water of Ham Lake, the wind hit them broadside, and they struggled against the waves. They quickly realized their best course would be striking out due south to the opposite shore and then turning head on into the wind as they steered toward their next portage.

By now the last two stragglers in the USFS group had arisen. Everyone was either having coffee and breakfast, or had already eaten. Several of them noticed the smoke plume from Steve's campsite and wondered about its size. But in due time it appeared to diminish, and then they noticed the canoe heading out of the inlet, and the group watched them struggle in the wind and chop. Seeing the canoe struggle convinced them they had made a good decision to keep their current location as a base camp. They started to gather their gear and some lunch for a day trip. They noticed the wind had picked up a little, and the sky had grown more overcast, so several of them packed their rain gear.

The USFS visitors did not yet know it, but there would be no rain in the Ham Lake region that day. The wind, however, which was already blowing at around 10 miles per hour, would continue growing and gusting throughout the day. From around ten in the morning to around ten that evening, the wind would blow anywhere from 10 to 22 miles per hour, with occasional gusts as high as 33 miles per hour. It was going to be blowy on the water and across the woods.

SEAGULL LAKE, MORNING—Layne was the first onto the water. He had broken camp often enough so that it took only minutes to finish the process. Because he was kayaking solo, all he had to worry about was his own gear and food.

Earlier, while Gus and Lee were still packing, he had pushed out into the water to have a brief morning paddle, check out the waves, and get a feel for the wind. So far, it wasn't too bad, but he could sense the growing warmth, and it felt to him, out on the water and farther away from the shore, that the wind was picking up. When he finally turned to head back to the campsite, he was maybe 100 yards out.

Overhead, there were a few scudding clouds, but from the feel of the wind Layne guessed those might well blow away by noon. He paddled back to the shoreline as Gus and Lee stowed their gear and climbed into their canoe.

"Let's make time," Gus said. He still hoped they could tour Three Mile Island's research plot, check out the oldest tree in Minnesota, and make it across the big water to the Alpine Lake portage before waves made it impassable.

The band of three paddled down the shoreline to the southwest. When they came around a small rocky outcrop, Lee pointed out his research plot, a good stand of black spruce with cedars down near the edge.

"There it is," Lee said, pointing to a wizened, thick-trunked cedar. As they paddled closer, they could see the large base of the tree, fastened into the shoreline's edge as though it was part rock.

"When the prescribed burn was going on five years ago," Lee said, "some of the locals came out here just to make sure this old tree wouldn't burn."

Lee explained that everyone knew about the tree, and the area's residents took pride in caring for their ancient neighbor.

They put ashore down from the ancient cedar and toured Lee's plot. When they moved away from the shore, they entered a stand of mature red pines. Some had been blown down, some were burned and dead, others were still living.

"This stand of trees germinated in 1801," Lee explained. "The smaller ones are only about fifteen inches in diameter. The larger ones were lost in the 1999 blowdown. The soil's shallow here, and the bigger, taller trees were vulnerable." Lee indicated examples of both.

"Here you have a two-centuries-old story. Bud Heinselman determined the date of that 1801 fire. There have been fires and of course blowdowns since. But some of these trees survived both the wind and fire."

The plot contained good examples of the different ways trees can succumb to fire.

"Here's an example of a tree that died from being girdled at its base by fire. To kill living tissues, the heat pulse has to go through the bark to the living tissues of the cambium underneath. Others can be killed by a crown scorch."

Lee showed them an example of a tree with all its pine needles at the top intact, but brown and clearly dead. "When there's a fire on the ground that's hot enough, the rising heat can be in excess of two hundred degrees. If the burn lasts long enough, it can kill the crown of a tree, which kills the tree. This tree was taken by the Forest Service's 2002 prescribed burn of Three Mile Island."

"But it didn't kill everything," Gus said.

"It usually doesn't. If there's a lot of wind, like today, then fires burn hot and fast and spread out in a kind of cone shape from a single point, jumping from treetop to treetop. That's a running crown fire. No conifer—except redwoods, which of course we don't have up here—can survive a running crown fire."

"When the top gets burnt, the tree's dead?" Layne said.

"That's right. If the top goes, so does the tree. But in order to get a running crown fire the elements have to be just right. It has to be dry, low humidity, and there has to be plenty of wind."

"Like now," Gus said.

"That's right. Like today. And of course there has to be a spark, something to ignite it. Up here that's usually lightning."

Lee showed them some examples of small black spruce trees that were part of this plot's understory. Some of them had died from the heat of the fire going through them, and some had survived.

"Wait until we see what the Cavity Lake fire did up in Alpine," Lee said. "For that fire, all the elements were right for a hot, intense fire, and it happened. If you remember, last July it was dry. The day was hot when it started. There was low relative humidity. And it was windy. But what that fire really fed on was the massive blowdown left over from 1999. It was thick, and it made the fire burn hot and intense—about the hottest I've seen it up here."

"How did the Cavity Lake fire start?"

"Lightning," Lee said.

They were looking forward to seeing the effects of that intense fire after they paddled into Alpine.

Down near the shore they passed the stand of ancient cedars.

"Why weren't these taken in any of those burns? In the 1801 fire that took those pines?" Layne asked.

"They're close to the lake. I'm sure their canopies were well hydrated. And they're set off from the rest of the forest. The soil here is really shallow and rocky, so there probably wasn't much on the ground for the fire to get close enough to these cedars to burn them. So they survived."

Lee paused. "If I'm right about this tree being a thousand years old," he said, "it's seen a lot of fires. Fires happen all the time up here. They have for the last ten thousand years, anyway. Probably longer."

The ancient cedar was magnificent. The familiar, dry ribbed bark made a gradual spiral, or twist, as it grew skyward, occasionally putting out limbs that appeared arthritic, like malformed arms with two or three gnarled fingers. But the top of the tree was healthy.

"It looks old," Gus said.

"It is old," Lee said.

"It's beautiful," Layne said.

The tree's canopy was fleshed out with thick, off-green garb.

BOB MONEHAN'S PLACE, SAG LAKE TRAIL—From Voyageur Canoe Outfitters, it wasn't much more than a long city block to Bob Monehan's place, farther up Sag Lake Trail. Bob was up early, admiring the sun and rising warmth. Apart from the wind, it was going to be beautiful . . . perfect for setting up his twenty-plus sprinkler water system.

Bob's sprinklers were fed by two pumps: one gas powered, one propane. They were both nestled into the rocks down to the right of Bob's dock, near the big intake hoses that ran into the lake. The big hoses ran up to the property and split off into smaller hoses that ran to the sprinklers. Some of the sprinklers were set out on the ground, some were on secure stands, and some—like the one on his fish house—rested squarely on the apex of the building's roof. All of them were connected by hoses that needed to be set up, unkinked, and tightened down. Wherever necessary, the line and sprinkler heads needed to be mended or replaced to eliminate leakage. The entire system had to be tested, including his gas- and propane-powered pumps. They also needed to be fueled. Gas-powered pumps will only run for around an hour. Propane-powered pumps, fueled by a 50-pound propane cylinder, will run for twenty-four hours.

Bob spent the morning working on his pumps, hoses, and sprinklers. He primed the pumps and started them, walking the lines, checking to make sure the sprinklers were operational, making repairs wherever needed.

It was early to be running the sprinkler systems, which normally weren't operational until residents could be sure temperatures wouldn't drop below freezing. Because of the problems that can happen from frozen water lines and systems, everyone waited until much later in the season to get their systems up and running. But Bob, like everyone else in the area, recognized this season was opening dry, hot (for this time of year), and windy. Bob knew fires could strike at any time, so he wanted to be ready.

Bob's system was based on designs he had seen elsewhere, including from his neighbors Michael Valentini and George Carlson.

THE VALENTINIS', SAG LAKE TRAIL—Michael and Sally Valentini were recent additions to the peninsula, having acquired their land next to Voyageur Canoe Outfitters on May 29, 1999, Sally's birthday. There was a one-room, rustic cabin on the property, to which they added a small bedroom and some essentials like electricity, DISH TV, and Internet but no running water.

When the Valentinis first bought their place, they were residents of Chisholm, three and a half hours by car across the Iron Range. They loved their place at the end of the Gunflint Trail. Michael had owned and run Valentini's Supper Club, the family restaurant in Chisholm, before getting out of the business. In 2006 he decided to move to their cabin permanently and start a one-man handyman business. Until Sally could retire, she visited two to three weekends a month, bringing him meals for the week. Sally was still working full-time in the Hibbing police department. But both were looking forward to the day Sally could retire and move from Chisholm to the Gunflint.

On this day, the Valentinis were in Chisholm at a family gathering. Michael had lost a cousin in a road accident on Highway 53, and every year his family gathered to clean up a section of the highway in memory of his fallen relative.

THE CARLSONS', POPLAR LAKE—George Carlson and his wife, Marilyn, built their cabin on Poplar Lake, near the midpoint of the Gunflint Trail, in 1986. George was used to talking with lots of people and doing both online and interview research. After the Windigo Lodge fire in 1991, George's own skill set was recruited to help start the area's first volunteer fire department. In fact, in that capacity he would discover something being done in Canada that—years later on May 5–7, 2007 (when combined with the participation of his then competitor Michael Valentini)—would have an inestimable effect on the lives of many of the cabin and resort owners residing at the end of the Gunflint Trail.

5 VOLUNTEERS AND WATER

The tragedy of the Windigo Lodge fire compelled the area's permanent residents—mostly resort owners—to contemplate a more robust firefighting approach than portable pumps and hoses. During the fall of 1991, while the tragic fire was still fresh in everyone's mind, there was an informal gathering at the Kerfoots' Gunflint Lodge. More than thirty people were in attendance, including many of the existing members of the Gunflint Trail Fire and Rescue Squad, as well as several other local permanent residents, all gathered to discuss what could be done.

Many along the Gunflint Trail were already familiar with the idea of a volunteer fire department. Tim Norman, USFS Fire Management Officer for the Tofte and Gunflint Ranger Districts, had long been arguing for more community involvement fighting fires. Tim and some of his firefighting colleagues often came up the Trail to present firefighting tips and techniques to lake homeowners associations and other groups. Since the early to mid-1980s, the spruce budworm had devastated some of the forest along the Trail. In particular, the forest contained lots of dead balsam firs, which some consider to be the most combustible and hazardous in a fire. Tim and his colleagues feared that if the area caught fire, the dead balsam and other fuel types could cause a major conflagration.

Volunteer fire departments are quasi-governmental organizations, and while their firefighters are all volunteer, the actual entity can be set up to have taxing authority. Many were in favor, but others did not want to take on the time, training, and commitment that such an organization required. Members of the Gunflint Trail Fire and Rescue Squad were familiar with the introductory and ongoing training required to be an emergency medical technician (EMT). Frankly, the idea of adding firefighting training to what was already required of them caused some members to balk at the idea. Also, the Rescue Squad was primarily comprised of friends and neighbors along the Trail who had known each other and worked together for many years. Unlike other areas of the country, in which resort owners sometimes considered other resort owners competitors, the Gunflint community had a more neighborly sense of place, in which friends helped friends, even though they might be competing for the same clientele.

Bruce and Sue Kerfoot had been members of the Rescue Squad for more than eleven years, and they—like other longtime squad members—had never required any formal organizational structure or designations of who would be chiefs, assistant chiefs, incident commanders, and more. With regard to creating a volunteer fire department, squad members also recognized that it was perhaps time for others to take over. Regardless if they continued as the Rescue Squad or metamorphosed into a volunteer fire department, they needed personnel and additional membership. The many new faces didn't have the decade-long experience of existing Rescue Squad members, but if a volunteer fire department could attract new blood and a greater sophistication into the mix, so much the better.

By the end of the meeting, the decision to move forward was supported by the majority, and the idea was passed. George Carlson, a resident on Poplar Lake, where the Windigo Lodge had resided, was elected to lead the effort.

Unlike many voting in the group that day, George was not a resort owner and had no role or business interest in the resort industry. He was a systems engineer and a headhunter. The name of his company was Targeted Quality Recruitment. His job was to locate viable candidates for open technology employment opportunities for a variety of clients. Eventually (1993–98) Motorola would be his exclusive client, and he would not only recruit for them but also help them craft their worldwide staffing strategy. And he did it all from his cabin on Poplar Lake, using phones, the Internet, faxes, and more. George was possibly the Gunflint Trail's first telecommuter.

Perhaps it was because George presented no competitive threat to the others, or because of his business background, or because he was an excellent researcher that he was chosen to lead the effort to help form the GTVFD. Regardless, one of his clear skill sets was going to be useful in creating the fledgling organization and, after its first six months, leading it as fire chief.

On June 9, 1992, an agreement with Cook County formalized the fledgling organization and identified where it would operate and its primary mission: "to provide fire protection to unorganized territory in the Gunflint Trail area." The agreement in part stated that the GTVFD was organized on May 26, 1992, "and now consists of at least 15 active fire fighters" with "firefighting equipment on permanent loan from the Minnesota Department of Natural Resources." More specifically:

> The Fire Department agrees to furnish fire service and fire
> protection to all of the property within the Gunflint Trail Fire
> District and further agrees that the Fire Department will make
> a reasonable effort to attend all fires within such Fire District
> whenever it is notified of such fires. The Fire Department agrees
> that it will maintain and house all of its fire apparatus without
> any expense to the County and will make all repairs of the fire
> equipment and apparatus at its own cost, and that it will furnish
> all the gas, oil, and other supplies needed to operate any fire
> equipment furnished by it, pursuant to this Agreement.

The agreement also spelled out a variety of other matters, including a $15,000 annual payment (at least for its first year) the county would pay to the organization for rendered services. The agreement also stated that the county "agrees to levy a tax against all the taxable property in said Gunflint Trail Fire District, as provided for in Minnesota Statues, Section 36.243, in a sufficient amount to raise the said $15,000.00 for June 9, 1992 through June 8, 1993, and subsequent years."

The Gunflint Trail Volunteer Fire Department was born.

George Carlson knew nothing about setting up a nonprofit or firefighting, so he began making calls. And he listened. He received assistance from a variety of sources. Tim Norman was more than happy to help George with a variety of tasks. It was Tim who first told George about one of the ways the Canadian government fought wildfires; they used portable, temporary sprinkler systems that sucked up nearby lake or river water (if a property was close enough) and spread a steady water pulse over the structures threatened by flames. Tim had slides of the water systems in use in Ontario, which he often shared during his meetings up the Gunflint Trail.

In late 1992, when George became the GTVFD fire chief, he remembered Tim's reference to the Canadian sprinklers and decided to find out more. Skilled at doing research over the phone, George first spoke with Canadian authorities and learned that these sprinkler systems were portable and temporary, using two-cycle, gas-powered engines. While the gas-powered, temporary pumps were attributed with preventing at least one billion dollars in structure damage, they had some reliability issues. The Canadian engines were capable of generating 300 pounds per square inch (PSI) of pressure, while the sprinklers attached to the systems operated at 30 to 70 PSI. To keep the engine's pressure low enough, they would have to run them at a low idle, which would sometimes cause them to falter and stop. Also, two-cycle engines required a gas-oil mixture to operate and tended to wear out faster than some alternative types of small engines. Finally, the time it took for these engines to run through gas was relatively short; if they couldn't operate long enough, they couldn't apply enough water to be maximally effective.

But when they worked, they worked well. And they were portable and temporary; the Canadian authorities could easily bring them in and set them up on an as-needed basis, providing they had enough forewarning and a nearby water source.

George was intrigued and interested enough to create some experimental systems on his own property and the property of a handful of neighbors. Along the Gunflint there were approximately 730 properties, containing thousands of structures. While temporary portable systems worked well in Canada, George wondered about creating permanent systems that could be maintained the way remote cabin and resort owners maintained their water systems. In the fall, when the threat of wildfire became minimal, these permanent systems could be blown out and drained ahead of the first frost. And in the spring, near the start of the wildfire season, they could be restarted, tested, and ready to operate in the event a structure was in the path of oncoming flames.

When designing these more permanent systems, he had several issues he needed to know more about. First, he needed to know how much water was required to protect structures from flames. George located and spoke with a structure protection specialist who told him if he could create a system that dropped the equivalent of a two-inch rainfall over a property (approximately 100,000 gallons for a typical cabin and surrounding area), the property should survive any fire. Considering the systems they were designing, it would take approximately twenty-four hours of continuous running to spread that much water. Once such a sprinkler system had operated for twenty-four hours, the structure would be safe from fire for approximately seven to ten days. That is, theoretically. The truth was, no sprinkler systems like those in Canada had been tried in the United States.

George also learned something about wildfire behavior. Wildfires are always in search of dry air and fuel. If they approach an area with humidity and moisture on one side and dry air and fuel on the other, they will move toward the dry, combustible fuel, even when being fanned by strong winds.

Unfortunately, in addition to the reliability and pressure issues that hobbled the gas-powered temporary engines used in Canada, they also did not run long enough.

George did some additional research and ended up speaking with the chief engineer of Briggs and Stratton, one of the world's foremost manufacturers of small engines. When the chief engineer heard about George's problem, he suggested using a four-cycle engine that ran on propane. A four-cycle engine was more reliable and operated at much lower PSI, and a 50-gallon propane cylinder could operate a sprinkler system for the necessary twenty-four hours. Propane was also managed using a closed fuel system, so no one would have to worry about hauling and pouring highly combustible gasoline into a gas tank in the middle of a fire.

No permanent systems like this had ever been created, so it took some experimentation to create a fully functional, operational solution. But after trial and error and using a variety of different kinds of equipment, George designed a system that worked. The first permanent wildfire sprinkler protection systems were created and installed in 1993. The approximately five or six systems George created (for himself and his neighbors) remained operational but untested by actual fire. Because the systems weren't cheap to install, there was no rush by Gunflint property owners to acquire them. Then on July 4, 1999, the infamous BWCAW blowdown happened, adding thirty million dead trees to the forest debris. If lightning struck or a camper's fire got out of control with that much fuel around, the subsequent wildfire could be tragic.

The blowdown was so significant that on July 28, 1999, President Bill Clinton declared Cook County a disaster area. Based on the declaration, George worked with Cook County to write the first FEMA Hazard Mitigation Grant in 2000. Under the grant, from 2000 to 2001, George and others working with him, including Michael Valentini, installed 188 sprinkler systems. The grant called for the government to cover 75 percent of the sprinklers' costs, while the remaining 25 percent was covered by the property owner.

Around this time Michael Valentini, whose rustic cabin was located at the end of the Gunflint, was trying to figure out how to move from Chisholm to Sag permanently. He made an effort to get to the end of the Gunflint as often as possible, doing handyman work and working as a subcontractor installing sprinkler systems. As part of the first FEMA grant in 2000, he was a subcontractor installer for George. During that summer he installed around twenty sprinkler systems, including two of his own, which helped him gain proficiency in designing, installing, and maintaining the systems. In fact, several of these systems were in his neighborhood at the end of the Gunflint.

On the morning of May 5, 2007, Michael Valentini, by this time a permanent resident of the area, and his wife were not at their cabin at the end of the Gunflint Trail. But by the next day he would be. And the Proms as well as several other property owners in the area would be glad he was.

6 POSNIAK STRIKES THE MATCH

The last day in the woods after a two-night camping trip is sometimes met with a mixture of melancholy and newfound appreciation for home-cooked meals and showers. Presumably after fixing and eating breakfast beside his morning fire, Steve settled in to review his newspaper one last time, perhaps while he enjoyed some morning coffee. While he read, he also would have heard the USFS people down the shoreline at the other campsite. The wind was in the right direction, the area between the site and his campsite was wide open, and sound travels a long distance over water. He probably heard them cleaning up after breakfast and getting ready for their day trip. In the late morning, around ten o'clock, the USFS group shoved their canoes into the water and set off for the portage into Cross Bay Lake.

If Steve had glanced up from reading to watch the four canoes, he would have seen them struggling in the wind. Once into the water, they quickly turned into the growing breeze, which would make the canoes easier to handle. Observing their struggle may have dissuaded him from taking one last paddle on the rough open water. He was probably glad he was heading back the way he entered, where the waterways were somewhat protected from the blow.

At some point Steve roused himself, deciding it was time to start gathering and packing his supplies. Much of his equipment was spread out over his campsite. No one had ever accused him of being a neatnik. It would take him a while to gather all his gear and get it back into the packs. And he still needed to burn his paper trash, so he had best get at it.

He was determined to burn all his paper. Technically, he knew he was supposed to carry it out. But it was much easier to simply burn it. It was a wood product, after all.

He must have gathered together his cooking utensils and started back toward his tent. On his way he could have picked up some of the other loose items scattered around the site. It shouldn't have taken more than twenty to thirty minutes to clean up his site and make sure all his loose supplies were stowed in a pile down near the green and brown canvas bags, where he could worry later about packing them.

At some point he returned to the fire grate, where there were still probably active coals around his fire that had escaped his heavy dousing. If so, he could have gathered together the coals around the edge of the fire ring and used some paper and kindling to bring the fire to life—at least enough of a small stick fire to burn his paper. Once the small flames were flickering, he fed in his trash, watching it flare up and then fold in on itself in blackened ash. Some of it only partially burned and remained around the edges of the fire ring. He would have to wait until it had cooled before grabbing it and feeding it into the main flames.

After kneeling for a while, Steve probably felt the need to stretch. He may have reached over and begun to dismantle his camp stove, inserting the different pieces into its nylon carrying sack. Once he had packed up his stove and emptied his coffee cup, he turned back toward his tent, which sat approximately twenty feet from the fire ring, down a small path through some cedars.

When Steve reached his tent, he bent down and crawled inside, starting to rummage through his gear, beginning to think about the best way to pack it. He had to take a moment to stow his camp stove in one of his packs and then begin tidying the inside of his tent, readying the bag and his other supplies for their eventual return to canvas. In spite of seeing how the others struggled on the open Ham Lake water, he may have wondered if he should try one last southern paddle before turning around and exiting the wilderness.

Not even Steve knew exactly how the fire started, but given the details we do know, the following is a likely scenario.

While Steve readied his gear and contemplated the rest of his day, a strong gust must have cut through the lake edge of trees, reached into the tepid fire, lifted some of the paper-light embers, and began carrying them on the breeze. The campsite's position, the spit of land, and the bordering trees caused the breeze to eddy back around the site. The circular wind carried and continued breathing life into a nearly burned ember, whether paper or bark it was unclear. The ember may have lifted over the rock, and when it came in behind the center stand of trees, it settled to the ground, still carrying a faint spark. It caught the edge of leaves and dried grass and passed its burn onto the tinder-dry debris. The grass caught and almost immediately flared, consuming some nearby leaves, and then moving on to the outskirt of pine needles beneath the black spruce boughs. Once it hit the dry pine needles, it fed like a starving child. And then another gust kicked up long enough to feed it an ample dose of oxygen, and suddenly it leaped into the undergrowth, climbing into nearby dry branches and dead saplings, spreading across the ground, feeding on pine needles, grass, and leaves. When the fire caught the first needles, it rapidly doubled in size and strength, continuing to spread, and then doubled again.

Eventually, Steve must have finished with his preparations in the tent and backed out on his hands and knees. Then he stood and started back along the narrow path through the cedars, up toward the fire ring and grill. When he passed through the narrow opening between cedars, what he saw must have struck him in his stomach like a fist. The branches and brush on the back side of the small island were on fire. There was more flame spreading away from the brush, consuming the dry grass and needles as it grew. He hurried around the trees, moving quickly toward the larger, more dangerous fire.

By now half the skirt of pine needles were aflame, and Steve must have jumped into the edge of it, kicking at it. But every time he kicked and stomped, the embers probably scattered like ground-born fireworks, tumbling away from his footfall. The breeze picked them up and spread them amoeba-like on a random path through the brush edge.

Steve raced back to his tent, rummaged through his gear for his small plastic container, and turned to the lake for water.

But the fire didn't pause. The flame fed on parched grass, pine needles, fallen tree parts, and leaves, continuing to satisfy its deep, primitive hunger. The breeze kept up a steady supply of oxygen, breathing the conflagration into substantive life.

Steve raced back to the flames. He splashed the water in a wide arc, and there was a sizzle and sudden cessation of fire in a few small sections. But the blaze was too wide and too dispersed, and most of it kept spreading and growing.

Steve probably stomped on the flames but with little effect. Then he turned and ran back for more water.

For the next thirty minutes, Steve Posniak exhausted himself in his futile attempt to douse the growing fire. Because his tent and supplies were on the leeward side of the flames, they were safe. But he wasn't thinking about his supplies. The wind came across the open water, pushed through the lakeside tree edge, fed the flames, and drove them into the upper branches of the black spruce. The fire spread to neighboring black spruce and devoured the dry under-cover of leaves, pine needles, and forest detritus until almost the entire woodland island was aflame. The wind was driving the fire wider and higher, and once it spread around the lower part of a tree's circumference, the breeze lifted some embers away from the tree's edge, carried them across to the opposite side of the campsite, and started the brush and trees beside the narrow canal on fire.

By then, Steve wasn't in a place to watch the fire catch, carry, and flare. He was still running back to the lake edge, filling his puny container, dousing the nearside flames, and returning again and again to the lake. But the fire barely noticed his efforts. The fire spread across his campsite, moving to its western outside borders, dispersing in front of the breeze in a growing giant amoeboid fan.

Steve must have finally decided his only recourse was to make sure the fire didn't spread due north into the open forest or somehow turn back and burn east and south. Now he just hoped the fire would continue burning to the northwest edge of the canal, burn down to the water, find nothing more to feed on, and finally smolder down to quiet, blackened ash.

The burning black spruce and small island of deciduous trees were sending a thin, steady column of smoke into the clear morning air. He must have wanted to take it all back. He must have wanted to keep the pillar of smoke from sending its signal into the billowing sky. His exertion and seeing his campsite in flames surely sickened him. It wasn't the kind of nausea that would cause his food to rise; it was heart sickness. He had stupidly started a fire and then left it. He must have felt awful about setting to flame his beloved wilderness, where he had so often come and camped and enjoyed what Minnesota's forest had to offer: solitude in a wild place. If he had the time to think about it, he must have felt embarrassed, saddened, ashamed, and afraid. But right now he was too damn tired to do anything but stand and watch and hope the fire would burn itself out, or that the visitors from down the shore were near enough to see the smoke and paddle back to Ham Lake to help him put it all out.

But the USFS people had long since paddled down the lake and across into a waterway that took them out of Ham Lake and away from the smoke and flame. By this time they were already one portage away and pretty far south along the meandering Cross Bay Lake waterway, too far away from Ham Lake to be able to return quickly and help.

Steve, exhausted, probably sat down on the granite outcrop, keeping vigil over the burning trees and brush near water's edge.

Within minutes the unimaginable happened. The wind picked up the embers beside the narrow waterway and carried them across twenty-five feet of water. The opposite bank contained dried grass, shrubs, and small pines. The parched forest took only a couple of minutes to embrace the embers, succor them to flame, and with an uptick in wind build a solid, fiery foothold that began burning to the west-northwest.

Too late, long after the point at which the flames could have been curtailed (provided there had ever been that point), Steve, more tired than he could ever remember being, ended his trips for water. He stood, almost certainly breathing heavily, sweating profusely in the cool morning, his forehead glistening, his body wet beneath his clothes, and stared at the runaway fire. His stomach probably churned, he felt dizzy, and he thought he was going to throw up.

Not a goddamn thing he could do.

At some point he must have pieced his way around the burning island of trees and brush to the rock escarpment and checked on his canoe. It was still pulled up on the sandy spit between two huge boulders. It was aluminum, so even if it was near the flames, it wouldn't have mattered.

Across the canal the trees and dry brush edge were aflame. The breeze coming in off the long sweep of lake was feeding it like an insatiable child.

Steve turned back to his tent and started pulling up stakes. It looked as though the path of the fire could be heading toward Tuscarora. The thought of the lodge and its buildings in danger surely made him crazy sick. He had no idea what to do, but knew he couldn't paddle back the way he had come. Within the next thirty to sixty minutes it was likely the fire would engulf the next portage over. He would have to pack up his supplies, fetch his canoe through the flames and smoldering embers, and make his way back to Tuscarora the best way he could. He just hoped and prayed the Tuscarora Lodge, house, cabins, and other outbuildings were out of harm's way, or that the wind shifted, or that someone somehow managed to turn it, or somehow managed to stop it. But for now, the wind was strong and appeared to be running straight toward that beautiful stand of pines on Tuscarora.

TUSCARORA LODGE, LATE MORNING—Andy Ahrendt liked working outside in the late morning. Earlier, Andy had called Jesse Derscheid, from Como Oil and Propane, and left a message about a possible propane leak under the crew cabin. When Jesse called back later in the morning, Sue asked that he come up as soon as he could. Now he was at Tuscarora, checking out the leak with Andy and making sure everything was operational and all their pilots were properly lit. He installed a new regulator, and now the two men paused in conversation.

Apart from the wind, it was a beautiful morning, and it was going to be another warm day. Andy appreciated the heat, but like everyone else in the region he hoped for rain. He could tell, from the slightly overcast sky, today was probably going to be another dry one, made all the more so because of the steady breeze. In fact, the long-term forecast didn't call for rain anytime soon. He knew next weekend's fishermen would appreciate it.

But like everyone, he was concerned about the possibility of fire. The forest was full of dry tinder, the USFS hadn't issued a fire ban, and next week more than a thousand fishermen were heading north to try their luck on the water.

Now on the hillside, having finished their work, Jesse and Andy paused. The conversation turned to motorcycles.

"The thing I like best about them," Jesse said, "is the way your surroundings are immediate. It's not like driving a car."

Andy agreed.

"When you're riding a bike and you go along the road, you can feel the changes in temperature, the wind. And you can smell things. You can smell things like that campfire."

Andy paused for a minute, sniffing the air. Suddenly he turned and said, "That's no campfire."

The two men hustled across Tuscarora to one of the area's highest points of land. It was a struggle to bushwhack to the top of the nearby hill. There were downed trees and debris from previous blowdowns, and even in the early spring it was tough going to reach the top. But when they did, they turned and looked out across Round Lake, watching a column of smoke rising in the distance.

"That's over by Ham Lake," Andy said. "I've got to call that in."

As Andy started off the hill, Jesse fished his radio off his belt. Jesse was a volunteer firefighter for the Hovland Fire Department, up Highway 61, northeast of Grand Marais. "I'll call the sheriff."

From the top of the hill Jesse kept watching the rising column of smoke, which appeared to grow thicker and more pronounced in the early morning air. When the sheriff's dispatcher came on, he introduced himself and mentioned he was a volunteer firefighter out of Hovland. "I'm up the Gunflint at Tuscarora, and there's a fire going on. We're watching it across woods toward Ham Lake. Looks like maybe less than a mile away," he said.

"Someone called two minutes ago," the dispatcher said. "We'll get somebody up there to check it out." She thanked him and signed off.

In the sheriff's office the dispatcher paused long enough to enter the call into the Cook County sheriff's log. She coded it as a call from the "Public" at "11:32." From her previous call, just two minutes earlier, she had already assigned this supposed wildfire a case number: 07161012. Now she entered the case number and typed: "Jesse from HOVFD reporting fire near Tuscarora Lodge–can see smoke and fire from across Round Lake."

Before she had a chance to call the local USFS, her phone lit up with more reports.

11:33—"Gunflint Pines reporting that Tuscarora had reported to them a fire across from Round Lake."

Gunflint Pines Resort and Campground was five miles up and across the Gunflint Trail from Tuscarora. Sue Ahrendt, Andy's wife and co-proprietor of Tuscarora, was already on the phone, running down her phone tree of people and offices to call in case of fire. Gunflint Pines was on her list because Bob Baker, the Gunflint Pines co-owner, was the assistant fire chief of the GTVFD.

At 11:34 a.m. the Cook County sheriff's dispatcher made another entry: "104 [Deputy Tim Weitz] copied re the fire and is responding with 118 [Deputy Duane Kuntz] from Brule Lake."

Finally, at 11:37 the Cook County dispatcher called the USFS dispatcher and relayed the information about the fire. "CC 10–5 info to USFS dispatch," read the log. Cook County dispatchers often used codes and numbers for shorthand entry into their logs. "CC" is Cook County, and "10–5" means "to relay or pass on information."

Back on the Tuscarora hillside it had been five minutes since Jesse Derscheid's radio call to the sheriff's office. In those five minutes Jesse watched the fire grow, which could now be seen clearly in the late morning. The wind was strong, the air smelled of smoke, and there was a rising column of white and gray above the flames. As he watched, the fire suddenly took on more life. The flames started hitting the tops of the distant black spruce. He watched them catch, then blow high into the air with flame, then jump to the next treetop and repeat. And it was coming on fast.

Jesse, familiar with fire because of the work he had done as a firefighter, knew this was a running crown fire. You cannot control a running crown fire. They are fast and dangerous, burning with unusual intensity, and he thought this one was headed straight for them. He watched it for a moment longer before realizing he had to get off this hill and move his truck to safety. He turned and started hurrying through the blowdown. He had to tell Andy and Sue it was coming on fast, and it was a running crown fire.

And it looked like it was heading straight for Tuscarora.

7

FIRST RESPONDERS

At around 11:35 a.m. Mike Prom, co-owner of Voyageur Canoe Outfitters, was working to get ready for Minnesota's fishing opener and the start of their summer season, when his phone rang. It was Shari Baker, co-owner of Gunflint Pines Resort, about fifteen miles down the road.

"Sue Ahrendt called from Tuscarora," Shari said. "She says they've got a definite fire."

Mike had been a volunteer firefighter for much of the time they had owned Voyageur. The current chief of the GTVFD was Dan Baumann. Unfortunately, Baumann was out of town. And since it was such a beautiful Saturday on the Gunflint, Shari's husband, Bob, assistant fire chief, had gone into the Boundary Waters on a brief fishing trip with their two sons.

The GTVFD had three fire stations along the Gunflint Trail; Seagull (Mike Prom's station), Gunflint, and Poplar. The nearest hall to Tuscarora was the Gunflint, just down the road. But Bob was in the woods.

"We'll go have a look," Mike said.

Mike had been answering these kinds of calls often enough to make him skeptical. Over the past two to three years Minnesota's Arrowhead region and Canada had experienced a noticeable uptick in the severity and frequency of wildfires. After the 1999 blowdown event, people were forewarned. The warnings were warranted, Mike knew, but those warnings—and subsequent wildfire events around the Gunflint and in Canada—made people jumpy.

Don Kufahl, also a volunteer fire fighter, worked for Mike at Voyageur. Mike and Don got into Mike's truck and headed to the GTVFD Seagull fire station, just a couple miles down the road. They didn't bother getting into their Nomex. First they wanted to make sure it was a wildfire and that it was something big enough to require their firefighting gear.

The previous year the Cavity Lake fire had burned just south and a little east of the Seagull station. The year before Cavity Lake the Alpine fire burned north of Seagull Lake. In the days and weeks after those fires were extinguished, people all over the region smelled phantom smoke and reported fires. Mike knew wildfires fed people's imaginations, because in their wake and when smoke drifted down from Ontario or Manitoba fires, he and his colleagues answered a lot of ghost calls. Although the wildfire signs on this day were troubling, it was still early in the season, and at the end of the Gunflint, there were still occasional patches of snow.

At the Seagull station Don and Mike fired up engine 583, a Type 6 fire engine. These fire engines are built on a Ford 550 chassis and look like pickup trucks on steroids. The Type 6 engine can carry three hundred gallons of water and comes complete with water pumps and hoses and an assortment of nozzles and related gear. Don and Mike pulled out of the Seagull station and started down the Trail. It was seven miles to the turnoff. They put the truck in high gear and barreled down the blacktop toward Tuscarora.

SEAGULL LAKE, MORNING—By the time Gus, Lee, and Layne returned to the shore from Lee's research plot, the view of Seagull's whitecaps building as the water opened up worried Gus. He still wanted to make it to Alpine, but he thought maybe they should first test the waters.

In the late morning Layne pushed out into the lake ahead of Gus and Lee. Layne was already well offshore while Lee was trying to carefully enter the bow of the canoe. When Lee pushed off a large boulder, he slipped. Suddenly he was in water up to the center of his belly. It was deep here, and he didn't want to capsize Gus so he hit bottom, soaking himself. The lake was so cold it sucked the breath out of him, and he scrambled back onto dry ground, the water running down his pant legs.

"You OK?" Gus said. It was hard not to smile at the awkward professor, his tan pants soaked to his waist and his shirt buttoned to the top, giving him the appearance of a drenched schoolboy.

"That water's f-f-freezing."

"We wouldn't last a minute in there."

"It wouldn't take long," Lee said, still trying to catch his breath.

"You want to get your stuff out, change your clothes?"

Lee thought about it for a minute. It was already warm enough out of the wind that he thought he might be OK. "If I stick them in my dry sack, they'll just get mildewy and no good for the rest of the trip."

Almost due west across the near section of Seagull Lake rose one of the lake's largest islands. On most maps it is unnamed, but the locals call it Eagle's Nest Island. It has a familiar, high rise Lee had climbed before. From the peak of granite, there is an excellent view of the surrounding country. And if it continued warming, the wind would act like a giant air blower. "Let's head over to Eagle's Nest. We can get a good view of what's going on over Seagull's big water, and it should be warm and windy enough to blow-dry my clothes."

Gus thought it was a good idea.

Lee sat down on a boulder and took a minute to get his boots off and empty them of excess water.

Layne was still out from shore in his kayak. He had paddled out forty to fifty yards and paused, drifting sideways, waiting. He wondered what was taking the pair so long. When he turned to consider them, he noticed an odd column in the sky, to the southeast. It was smoke, maybe from a nearby campfire. He began to paddle back to his companions when he finally saw them get into their canoe and start toward him.

"Hey!" he yelled back to them. "Check it out. I guess we're not the only ones camping on the lake."

Gus and Lee looked up to where Layne pointed and noticed the column of smoke.

"I hope they're being careful," Lee said.

From this perspective, they could not determine the distance of the smoke column. Layne thought there were maybe campers on the other side of Three Mile Island.

"Lee fell in and got soaked," Gus yelled over to Layne. "Let's head over to Eagle's Nest," he pointed to the rocky island due west. "It's getting clear and warm enough we can hike up on that ridge and see what it's like out over Seagull. And Lee can dry out."

Layne nodded and quickly turned his kayak around, paddling toward the large island.

For the next several minutes he stayed well ahead of the canoe. Even though they were leeward of Three Mile Island, the wind was still strong here, and the waves were bigger. Not whitecaps, but Layne guessed on the other side of Eagle's Nest, where the water was open all the way to the Alpine Lake portage, it was going to be rough. He turned around to wait for Lee and Gus in the canoe and was surprised by what he saw. He waited a few more minutes for Gus and Lee to catch up.

"You guys gotta turn around and have a look at this," Layne said.

Gus dug in, and Lee helped him turn the canoe so they could see into the southeastern sky. The smoke column had doubled in size, clearly prominent, and from this vantage point it appeared to be several miles away.

"That has to be a wildfire," Lee said.

"Let's get up on that ridge, where we can have a better look," Gus said.

There was a narrow stretch of water between the peninsula on the eastern edge of Eagle's Nest and a very small island further east.

The three paddled through the opening and turned west along the peninsula, coming in behind the protected shore where the water was calmer. Down in the center of the cove they beached the canoe and kayak. It was a climb to get to the top, where a high granite promontory marked the southeast side of the island. By the time they reached it, the wind seemed even stronger than before, unencumbered by land, blowing straight and hard. In the several minutes it had taken to get there, the column of smoke had grown more pronounced.

"That's a fire," Lee said. "That's a damn big fire."

While they were serious, it was another occasion for Gus to smile. They were on the edge of wilderness, witnessing one of the Boundary Water's seminal events, and he expected one of the world's foremost forest fire experts to come up with something more profound than "that's a damn big fire." But then Gus guessed the professor's understated observation was probably in direct proportion to what he knew—based on years of research and observation in the field—would be the probable impact and growth of this fire.

The three stared at the growing column of smoke, beginning to think their day (and trip) was about to take a radical turn.

USFS, GRAND MARAIS OFFICE—At 11:38 a.m. Kurt Schierenbeck answered the phone and heard Sue Ahrendt, the co-owner of the Tuscarora Lodge, on the other end of the line. She told him there was a fire off Ham Lake and it looked serious—at least serious enough so she was calling everyone on her phone tree.

"We'll get someone up there right away," Kurt said.

As the fire duty officer for the day, Kurt fielded calls and coordinated the response. As soon as Kurt hung up the phone, Vance Hazelton, the Type IV IC for the day, walked into his office.

"Get your team together," Kurt said. "You need to head up to Tuscarora Lodge. There's a report of a wildfire, and this one sounds like the real deal. I'll start making more calls."

"We're on it," Vance said.

As the Type IV IC, Vance would be orchestrating their response on the ground, if it was needed. He turned, already thinking about his team and equipment. From Grand Marais it was a forty-seven-mile drive to the Tuscarora turnoff. If he wanted to get up there within the hour, he would have to hustle.

Vance assembled a crew of six, two riding with him in his USFS pickup, and three following in a Type 6 engine being driven by USFS firefighter Pete Lindgren. Pete was also a Type IV IC. The year before he had briefly assumed command of the Cavity Lake fire, before it grew too large for his span of control and he requested a Type III IC. But today, Vance was the on-duty Type IV, and in the few minutes following Kurt's order he briefed everyone on where they were headed and what to expect, and told them to get their gear. By 11:58 Vance's firefighters had donned their Nomex, tossed their packs and related gear into the trucks, and were pulling away from the station in their two trucks, Vance out in front.

As Vance headed up the road, Kurt gave him another call, letting him know engines and more personnel were right behind him and that he had called in aerial support. Cell phone reception up the Gunflint Trail was nonexistent, so they agreed to keep in touch using their USFS radios.

At 12:10 p.m. Kurt made a call to USFS dispatch to let them know what was happening.

"Vance Hazelton will be the incident commander for the fire," Kurt said. Once the fire response was under way, Kurt's focus intensified. He had a lot more calls to make, so he kept to the essentials. "We've dispatched a CL-215 and a Beaver for the air attack."

CL-215s are large yellow planes that can carry and drop 1,300 gallons of water over affected areas. Once empty, they can drop down onto lakes and refill their tanks by scooping water on the fly.

The on-call air attack team rotates between each of Minnesota DNR's three air tanker bases: Hibbing, Bemidji, and Brainerd. These three tanker bases usually have at least one CL-215 and a light helicopter. On this day the Hibbing base was on call, so it also had a fixed-wing aircraft—a Beechcraft Queen Air—which would be used to take an air attack supervisor up over the flames.

Once Kurt made the call to USFS dispatch in Grand Rapids, they reviewed it and took Kurt's request down the hall to the Minnesota Interagency Fire Center (MIFC). MIFC is responsible for locating and dispatching all the fire resources, including air support. On this day MIFC determined that of the three tanker bases on duty, the Hibbing base was closest, and they contacted Jody Leidholm, the on-call air attack supervisor at the Hibbing base.

This was Jody's eighteenth season on fire. Jody called his pilot for their Beechcraft Queen Air. The pilot would get him into the air immediately. Jody needed to be at least five to ten minutes ahead of the CL-215, because he would be acting as an air traffic controller over the fire. He needed to arrive first, review the scene, and plan the attack—possibly calling up additional resources. He was also the extremely important eyes in the sky for the team of firefighters on the ground.

Within minutes of getting the call from MIFC dispatch, Jody and his pilot were in the air, with the CL-215 close behind.

The Beaver was a De Havilland DHC-2, which many consider to be the best plane available for getting into the bush, particularly when the bush contains lakes and rivers. The Beaver has pontoons that enable it to easily land and take off on almost any lake in any condition. During fire season there is almost always a Beaver in the air, searching the boreal forests for any sign of fire. These planes also fly search and rescue missions. The Beavers are part of the USFS air support, based out of Ely. Once Kurt let USFS dispatch know he needed a Beaver, they let MIFC know, and the Beaver was called up.

Because it was midday during the fire season, a plane was already in the air, scouting the entire area on a standard protocol reconnaissance flight.

At the USFS offices, Kurt continued running down his resource and contact list. He contacted the USFS dispatch office in Grand Rapids to begin requisitioning the equipment and other resources he would need. Kurt knew USFS dispatch would review the order, augment it if necessary, and then take it down the hallway to the MIFC dispatch, who would be responsible for ordering the resources.

After getting the key equipment called up, he started making calls to firefighting colleagues, just to let them know a fire might be brewing. He called the DNR, the Bureau of Indian Affairs (BIA) in Grand Portage, and others. All of these governmental entities have people and equipment that can be called upon if the circumstances require it. Kurt's initial call was just informational, to let them know they should be on standby in case they were dealing with something serious.

FIRE ENGINE 583 ON THE GUNFLINT TRAIL—At 11:44 a.m. Mike Prom radioed Cook County dispatch and let them know he and Don Kufahl were in Fire Engine 583 headed for the Round Lake Road turnoff (to Tuscarora Lodge).

The dispatcher thanked Mike and entered the time, the entity making the call, "GTVFD," and a brief description. "Prom will be standing in for Bauman as he is out of town—he and another fireman [Don Kufahl] will be going to Cross Landing to evaluate the situation."

By 11:49 Mike and Don were approaching the turnoff to the Tuscarora Lodge, where Jesse Derscheid's Como Oil and Propane truck was parked beside the road. Jesse was out of his truck, looking up toward Round Lake, which was increasingly covered by a smoky haze. The wind was strong, and when Mike and Don passed, Mike rolled down his window.

"It's pretty bad, and it's coming on really fast," Jesse said. "It's a running crown."

"We'll go have a look."

"I think everybody should get out of there."

Mike nodded. "I agree," he said, and coasted by Jesse, continuing up the road.

They saw plenty of smoke rising in a windy sky, much of it already hazing across the area ahead of them.

"That does not look good," Mike said.

"Looks like it's bearing right down on them."

"I'm calling in an evac order," Mike said.

"Good idea."

The GTVFD radio frequencies had access to the Cook County sheriff's frequency. During fire events volunteer firefighters and the sheriff were responsible for evacuations and structure protection. The sheriff was responsible for issuing evacuation orders. But everything was happening too fast.

Mike and Don were both judging the fire by the smoke, which was thickening as they continued driving. By the time they passed the halfway point, they saw spotting on the left side of the road.

Fire spotting occurs when there is heavy wind pushing a fire, and burning debris is picked up and thrown out in front of the primary flames. The embers set down, and if they find purchase in dried leaves or twigs or some other kind of tinder, a small spot fire flares up. At first the spot fire can be small, but depending on the conditions and available oxygen, it can quickly grow into a substantial fire of its own.

When Mike and Don saw spotting in the forest to their left, they knew it was a bad sign. Not only did it tell them the winds were strong and fanning the flames, but now they knew they were directly in the fire's path. Normally, running crown wildfires spread in a widening V shape as they move out from the point where the fire began. In this instance, the fire was being quickly fanned by high winds, moving it in a narrower, cigar-like path. And judging from the smoke, wind, and the spot fires well out in front of it, it was making rapid progress in their direction.

The pair of firefighters also noticed the flames' color. When fires in the tops of trees are burning full force and hot, they consume a lot of oxygen, starving the spot fires in front of the blaze. The color of those spotting flames was a faint orange, like looking through a camera filter, Mike thought. Mike had seen the phenomenon before and knew that once more oxygen became available, these hot spots would flare up like mini-fire explosions. They did not want to be there when the oxygen caught up with the flames.

The pair drove up the road, paying close attention to the occasional spotting out in the forest to the left. As they rounded the last curve toward Tuscarora, trying to move as quickly as possible, they saw Sue Ahrendt and one of her employees approaching in Sue's gray Suburban. They had a canoe and were heading down to the Cross Bay Lake entry point. Sue slowed as the big Type 6 fire truck passed. Mike slowed his truck to barely a crawl and rolled down his window.

"Where you going?" Mike said.

"We've got a guy solo out by Ham Lake," Sue said.

"You're going in after him?"

They were both still moving but very slowly. "He's alone. What else can I do? We've got to try."

As an outfitter and resort owner himself, Mike understood. "OK," he said.

But the moment Sue and her assistant passed, Mike thought, *They shouldn't go in. I should have stopped her.* He should have told her to keep driving down Round Lake road to the Gunflint.

Over to the southeast, with the flames coming on fast, it was too dangerous. But she was already past, and Mike knew there were more people at the lodge. There wasn't enough time to turn back.

When they reached the road in front of the Ahrendts' house, the smoky haze was thick, and now it was just a matter of time. Smoke like this presaged heavy burning, and they did not want to be here when the flames finally burst through the haze.

Both firefighters got out of the truck and headed up to the Ahrendts' front door.

Inside, the Ahrendts' ten-year-old son, Daniel, and twelve-year-old daughter, Shelby, were trying to figure out what to pack, having been told by Sue and then Andy to gather everything they thought they needed or wanted—whatever they thought was most important to save from the flames. Mike and Don both offered whatever assistance they could, reminding Andy they needed to move fast.

The Ahrendts' black Labrador retriever, Denali, was wondering about the haze and commotion.

Finally, Andy gathered his son, daughter, Denali, and some articles—whatever he could pick up that he knew he needed to save from the fire—and got them into his Jeep. The air continued to thicken with haze and smoke. Andy put the jeep in gear and started down the road. As they broke around the first curve, the brush and trees on both sides of the road were on fire, and they headed straight into the tunnel through the flames.

Off to the right, at the Cross Bay Lake entry point, they saw Sue Ahrendt's gray Suburban in the narrow backwoods parking area.

At 12:12 Mike, on his way down the road, radioed Cook County to let them know they were in the process of evacuating Tuscarora.

It had been more than ten minutes since Mike and Don had passed Sue and her coworker, who were headed in to find, help, or rescue Steve Posniak. Sue worried about their lone, overweight camper. She knew the fire had come on so fast that if he was on the edge of Ham Lake, he could be in harm's way.

Sue and her coworker paddled across the small bay of water. Smoke and haze was thickening the air. Sue knew the second they saw fire, she had to turn around. She wasn't about to risk their lives for the camper she guessed had very likely already paddled to safety.

They crossed the water at full speed. Overhead the wind was blowing hard, and the air was thick with smoke. They beached the canoe, and Sue told her assistant to wait while she started up the first, long portage into the next small waterway. Near the top of the portage she could hear and see the rapid approach of fire, coming straight at them, hopping the treetops like yellow flaming acrobats. It sounded like a freight train. The oncoming conflagration was enough to convince her they had to abandon their effort. She rushed back to their canoe, where they hustled back into the water and hurried across the small waterway, returning to the Cross Bay Lake entry point. The pair piled into Sue's gray Suburban at about the same time the nearby woods started to ignite.

Sue drove down the entry point exit and pulled out onto Round Lake road at the moment Andy's Jeep was coming out of the fire, down Round Lake road. The Type 6 fire engine followed behind Andy's Jeep, and the three vehicles passed through the flames to safety. Within minutes they were at the Gunflint Trail turnoff, their vehicles pulled to the side of the road. The weather was windy but warm.

At 12:15 p.m. Mike radioed Cook County dispatch. "Where's Sheriff Falk? We've evacuated Tuscarora. They think they have an overweight camper on Ham Lake. Can you relay that to the USFS?"

Dispatch said they would relay the information to the USFS and see if they could locate the lone camper from the air.

DNR air attack supervisor Jody Leidholm was in the passenger seat of the Beechcraft Queen Air. As soon as he got overhead, he heard the request to look for the camper, and he thought he had found him, right off the point on the westernmost campsite of Ham Lake. He saw a lone camper wearing dark blue with two packs in an aluminum canoe. He spent a moment to call it in and then turned his attention to the runaway blaze.

What he saw was alarming.

His CL-215 would soon be on the scene, so he figured out its first plan of attack and radioed the pilot to let him know where to drop.

"Tanker 266 this is Air Attack One," Jody said.

"Go ahead Air Attack One," the pilot responded.

Because Tuscarora Lodge was in the path of the fire, Jody knew they needed to start by dropping immediately on Tuscarora and its cabins and outbuildings.

"Start dropping alongside the Tuscarora Lodge and its outbuildings." Full salvo loads unless told otherwise.

Once he instructed the CL-215 to start there, he called dispatch to request additional resources.

"Superior Dispatch, this is Air Attack One on Superior Net."

Jody followed the standard radio protocol. First he called the nearest dispatch office. Up on the Gunflint Trail, reception can be spotty. From experience Jody knew his best bet was Superior dispatch. Once he called them, he identified himself—Air Attack One. Then he let them know the radio frequency he was on, so they could use that frequency to respond. It was 12:45 p.m.

"Go ahead Air Attack One."

"We've got twenty-five to forty acres of a torching, running crown fire heading straight for Tuscarora Lodge. It's about a quarter mile out and burning fast. There are several structures in danger."

Jody knew what dispatch needed to hear. From his quick reconnaissance, he also knew they were going to need all the air attack support they could access. On this day there were two CL-215s under DNR contract in Bemidji and another CL-215 in Brainerd.

"I need another CL-215," Jody said.

Dispatch confirmed they would call up Brainerd and signed off.

Within minutes Jody examined the fire and the threat to structures and decided to call up Tanker 21, the P-3 Orion from the Ely tanker base to assist in protecting the resort structures. The P-3 is designed to drop three thousand gallons of red fire retardant onto the fire. The retardant is more commonly called mud; it is a mixture of water and chemicals designed to retard the growth of flames through vegetation. You do not want to be under a mud drop, either on the ground or in a vehicle. The red retardant is heavy enough to drop a man or smash a windshield.

Thankfully, the second CL-215 out of Brainerd was in the air moments after Jody requested it. Unfortunately, it was raining and lunchtime in Bemidji. Normally, pilots during fire season are restricted to their bases. But unlike on the Gunflint Trail, in Bemidji there was a steady rain. Thinking the likelihood of a fire with this much precipitation was unlikely, the four Bemidji CL-215 pilots requested a leave to go to lunch in town, and dispatch approved it.

Now they were recalled and rushed back to their base.

• • •

On the ground, everyone at the Gunflint Trail turnoff to Tuscarora got out of their vehicles and looked back at the smoke and haze behind them, nearly a mile away. They saw the first CL-215 fly overhead, starting to attack the flames. As soon as it arrived, it began spreading water on the flames. Not long after the first CL-215 arrived, a second was in the air, the pair of them trying to create a defensive arc around the Tuscarora Lodge and buildings, setting down a firebreak to save the place.

Now that they were out of harm's way, Don and Mike realized it was only a matter of time before they would have to once again get close to the fire. They fetched their Nomex gear out of Engine 583 and began to pull it on.

8 A GROWING FIREFIGHT

On Saturday, May 5, Sheriff Mark Falk was in his garage, enjoying his day off, working on his 1967 Suzuki motorcycle. In the middle of Sheriff Falk's tinkering, the first call regarding the fire came in to him at 11:40 a.m., ten minutes after it was first reported.

After two decades in law enforcement, the past two years as sheriff, Mark understood the needs of his office. He started in law enforcement with the Ortonville police department in 1987. Two years later he joined the Grand Marais police force, where he worked until 1991, when then Cook County Sheriff John Lyght hired him as a deputy. Lyght was succeeded as sheriff by Dave Wirt, who hired Mark to be his chief deputy in 1995. In 1999 and 2000, Mark spent nearly a year as chief of police for Canby, Minnesota, before returning to Cook County as a communications supervisor and jail administrator, and finally as deputy. Then in 2002 Sheriff Wirt reappointed him as chief deputy. When Wirt retired in February 2005, Mark succeeded him as the new Cook County sheriff.

Having worked in Cook County and in the sheriff's office for much of his career, Mark was familiar not only with the requirements of the job but especially with the sheriff's complicated responsibilities when dealing with wildfires. In August 1995, his first year as chief deputy, he worked the Sag Lake Corridor fire, which burned nearly thirteen thousand acres on Saganaga Lake. That fire threatened many cabins, resorts, and other structures at the end of the Gunflint Trail.

His first year in office as sheriff involved a literal trial by fire when the Alpine Lake fire broke out in August 2005. The Alpine Lake fire was the first to occur in the BWCAW since the 1999 blowdown. It burned more than one thousand acres and threatened nearby cabins and resorts before it was ultimately contained.

Then on July 14, 2006, at the height of the tourist season, the Cavity Lake fire broke out. Because of the time of the year, its location (near people and private property), and the weather and site conditions, fire managers immediately began to try to suppress the fire. On the first day of the Cavity Lake fire, air support dropped sixty thousand gallons of water on the flames.

Then one month later, in mid-September 2006, several wildfires broke out in the Gunflint District. Known as the East Zone Complex fires, they also threatened people and structures and before they were contained, burned approximately six thousand acres.

Thankfully, none of the preceding fires had resulted in loss of life or personal property. In fact, there were several positive outcomes from the blowdown and the subsequent wildfires on which Mark had already worked—as both chief deputy and sheriff.

First, in the wake of the 1999 blowdown, when everyone recognized the potential fire danger posed by the BWCAW's massive increase in fuels, Cook County, the sheriff's department, the DNR, the USFS, and others worked on and finalized the *Cook County Emergency Operations Plan* and another document Mark recalled as *The Northeast Minnesota Emergency Response Plan*. When the Cavity Lake fire started, all the relevant team members, including Sheriff Falk, used those plans to help manage the fire. The plans had also familiarized him with all of the key governmental groups and individuals and their responsibilities when working on a wildfire. The Cavity Lake fire highlighted the value of both plans, so much so, that even if there were no wildfires happening, Sheriff Falk recounted, "We would meet one or two times a year just to dust them off and review them and make any necessary corrections or revisions to them." The plans became an integral part of the team's functioning, particularly when making decisions on how best to manage wildfires.

Even without the periodic review of the documents, the Alpine Lake, Cavity Lake, and East Zone Complex fires all gave Sheriff Falk and others plenty of practice both using the documents and getting to know what needed to happen when and by whom.

The "by whom" was particularly important. Documents like these can be stale, dry, and tedious. But when names, faces, and personalities are connected to key players outlined in the plans, they take on a life of their own. One of the most important outcomes of the past three fires since Alpine Lake in August 2005 was a familiarity with most or all of the players who would need to be involved in this growing wildfire.

The sheriff's office was responsible for evacuations and law enforcement in the area. If any structures or areas with cabins or resorts were threatened, the sheriff had to make the hard call about an evacuation. And of course, depending upon the size of the evacuation, the sheriff and his deputies needed help to get the word out and let people know they needed to evacuate. There were lots of details to consider, many of them emotional. Anyone who is being evacuated from his or her home has plenty to think about.

Since the previous three fires were still fresh in Sheriff Falk's mind, he knew his duty was to head up the Trail as soon as possible. And even he knew that with the unseasonable warmth, dryness, and wind, this could be a wildfire that would get out of control fast.

At 12:25 p.m. he radioed dispatch to make sure they had alerted his colleague Nancy Koss, the Cook County emergency management director. The director would need to know about this growing potential problem, too. Dispatch confirmed they had.

Then at 12:26 dispatch let Mike Prom know Sheriff Falk was headed up the Trail and would contact him when he got near the fire.

• • •

Back on the Gunflint, Vance Hazelton, the Type IV IC—who would be in charge of the fire once he finally reached it—and his team of firefighters were approaching in Vance's truck. The USFS fire engine followed close behind. By the time they reached the Gunflint's halfway point, Vance could see a widening column of smoke rising in the distance. On their way out of Grand Marais, the skies were actually overcast, and at one point a fine mist dropped from the clouds. They actually had to run their windshield wipers. But the rain was only temporary and barely wetted the ground.

As they neared the fire, all the firefighters began seeing what they recognized as the signs of a growing, significant wildfire. The USFS radio chatter became more cacophonous as the column of smoke widened.

Minutes earlier, Tom Lynch from the DNR had been monitoring scanners at the DNR office in Grand Marais when he picked up the first chatter regarding the fire. The scanners can be difficult to hear and don't always provide important details, so Tom called Cook County dispatch for more information. As soon as he found out the location of the fire, Tuscarora, and reflected on his morning conversation with Kurt, when both agreed the warmth, low relative humidity, and wind were unfavorable conditions, he knew he should head up to assist.

Tom had been working in fire since 1994, when he was first employed by the DNR as a smoke chaser. Since then he had worked on and off for the DNR and other organizations in a variety of firefighting and fire mitigation and training roles. Coincidentally, just three months earlier, he was part of a firefighting training class in Grand Rapids, Minnesota, that used Tuscarora and the surrounding forest as the scene of a mock wildfire, using maps and the location to train others on how such a fire could be addressed. Tom also lived midway up the Gunflint Trail, so he was very familiar with the area. Also, like

Kurt, Tom had worked in fire long enough and had enough experience and class work to have completed his Type III IC certification.

From what Tom now heard about the origins of this fire, he understood the USFS would be taking the lead to fight this fire. But he recognized the conditions were not good for the start of a wildfire, so he and a fellow firefighter hurried into a Type 6 engine and begin barreling up the Gunflint Trail just minutes behind Vance and Pete.

Type IV ICs have enough training to manage a small coterie of personnel and equipment. As they sped up the Gunflint, Vance kept glimpsing the column of smoke, beginning to worry. Still several miles short of the Tuscarora turnoff, he was guessing the number of resources that should be called out on a fire this size already exceeded his span of control. And the signs were getting worse. The wind was up, and in spite of that little bizarre misting that had happened down the road, the overhead clouds didn't look like they would be giving up any moisture. It was bone dry and getting warmer.

USFS radios have numerous channels that can be used. The formal channel uses a repeater to boost calls from far up the Gunflint to the USFS office in Grand Marais. USFS firefighters also had access to a tactical channel, more often referred to as a tac channel. Tac channels were good for very short distances, providing there were no big obstructions between the sender and receiver.

Vance, occasionally glimpsing the rising, widening column of smoke up ahead, decided to radio Pete Lindgren on the tac channel in the Type 6 engine behind him. He remembered that Pete for a very short time had been the Type IV IC responding to the Cavity Lake fire. And since that fire quickly accelerated up the IC type chain, he wondered *when* Pete had decided to make the call to move it from a Type IV to a Type III fire. Each wildfire had a life of its own, and the decision to elevate the response to one was never a clear-cut choice. Vance was pretty sure this one already qualified for a Type III response, but he wanted corroboration from someone who had already done it.

Vance hailed Pete on the tac channel. Once they were connected, Vance said, "When you were on the Cavity Lake fire, when did you elevate it to a Type III?"

Pete knew what Vance was asking. He had been watching the thickening column of smoke as long as Vance. "I probably would have called it by now," he said.

Vance concurred.

At around 12:30 p.m. Vance used the official channel to call USFS dispatch. He was still ten minutes away from the Tuscarora turnoff. "This is Vance Hazelton, and I'm requesting the response to the Ham Lake fire be elevated to a Type III. Immediately."

Dispatch received his request and told him they would pass it on.

Kurt Schierenbeck was still the on-duty fire officer in the USFS Grand Marais office. For the past hour he had been busy calling up equipment, including the air attack, and monitoring the increased radio traffic on the line. He was keeping in touch with Vance and his team, and making sure other agencies up and down the North Shore and in the Arrowhead region and Ontario were alerted to the blaze. When dispatch told him Vance had requested the wildfire be elevated to a Type III incident, Kurt agreed and began thinking about the proper way to get on top of this fire and assume command. The reports they were getting from near the fire and from the sheriff's dispatch all pointed to a serious and growing fire. One of his first acts was to find a replacement for his duties in the office, so he could head up the Trail. Within minutes he had radioed the Beaver plane, already in the air, and asked the pilot to pick him up at the Devil Track Lake seaplane dock. Kurt needed to have a better look at the fire he had just decided to take on.

At 11:34 a.m., just four minutes after the first call reporting the fire came into Cook County dispatch, Deputies Tim Weitz and Duane Kunz were in a patrol car near the Brule Lake landing. Brule Lake is just twenty-odd miles south and a little east of Tuscarora, at least as the common raven flies. But much of the distance between Brule and Round Lakes is wilderness. If you know the back roads to get from Brule to the Gunflint, the distance is a little more than fifty miles. The day was quiet near Brule Lake. Tim and Duane decided they had better head up to the fire scene, just in case. They radioed Cook County dispatch and told them they were on their way.

Since some of the back roads to get from Brule to the Gunflint Trail are unpaved, it took the pair of deputies a little more than an hour to reach the turnoff for the Tuscarora Lodge. When they approached the spot, they saw a growing assemblage of emergency vehicles, firefighters, and civilians. There were four fire engines—one from the DNR, the engine commandeered by Don and Mike from the GTVFD Seagull station, another from the GTVFD Poplar station, and the USFS engine. The Ahrendts' Suburban and Jeep were beside the road, with Andy, Sue, their two kids, and Tuscarora's two workers. Jesse Derscheid with his Como Oil and Propane truck was still there, with several others. Don and Mike were out of their truck, looking at a map. Vance Hazelton and his crew had also just arrived. And while everyone was worried sick about what was happening up at the Tuscarora Lodge and outbuildings, down here by the road it was relatively smoke free. Other than the wind, it was a cloudy but beautiful early May day. But no one was thinking about the pleasant weather.

Once Vance Hazelton arrived at the Tuscarora turnoff, he created a unified command with Mike Prom. Everyone reiterated that Mike and now Tim Weitz as the sheriff's office representative were now responsible for evacuations and structure protection, while Vance and his USFS team, including the planes, were responsible for fighting the wildfire on the ground.

Air Attack Group Supervisor Jody Leidholm was still overhead, keeping tabs on the fire from the passenger seat of his Beechcraft Queen Air. The CL-215s had been working Jody's plan and were making good headway against the flames. Both aircraft were now working to paint an arc of water around the perimeter of Tuscarora.

From the air Jody was also in touch with Vance on the ground, sharing his perspective of the fire from his bird's-eye view.

Once a USFS incident commander comes onto the scene, they begin working through a checklist of items that need to be considered and addressed. Many of these are obvious, involving communications and logistics, determining if structures or people are threatened, if anyone has been injured, or if buildings have already succumbed to flames. As part of this checklist, one of the first tasks of an incident commander is to give the incident a name. In the USFS, fire names are based on the geographic location near or at the place it started. Off the Gunflint, the practice has usually involved lakes: the Alpine Lake fire, the Cavity Lake fire, and so on.

Jesse Derscheid and Andy Ahrendt, as well as Jody in the air, had already seen the point of origin of this fire and conveyed that information to others at the scene. Following his fire naming guidelines, Vance gave the sudden and growing wildfire a name: the *Ham Lake fire*.

Having received a full briefing from Mike Prom, Vance and some of his crew were reviewing an area map, continuing to keep in contact with Jody and Kurt. An Orion P-3 tanker joined the other planes fighting the fire.

The heavy smoke covering the Tuscarora Lodge was obscuring the scene from the air and the 215s, and now the P-3 needed guidance on whether everyone was out of the area and what the others had seen on the ground. The spot fires and heavy smoke were also obscuring the front line of the fire.

Tom Lynch and his fellow DNR firefighter remembered Forest Road 1495, which turned off Round Lake Road and bent around to the western side of Round Lake. Because he thought that road could get them in front of the fire, they drove in and had a look. But the spot fires and smoke were troublesome here, too, so they headed back to the Round Lake Road turnoff on the Gunflint Trail.

Vance also told everyone the fire had been elevated to a Type III incident, and that the Beaver was headed to Devil Track Lake land-ing to pick up Kurt Schierenbeck, the Type III IC. Kurt would fly over and have a bird's-eye view of the fire and figure out next steps.

Meanwhile, the aerial bombardment with the red fire retardant and water was starting to have an effect (from what the pilots could see from above). The 215s and the P-3 had begun creating a fire-break between the woods and Tuscarora's buildings. As the wildfire approached the edge of that break, the wide band of red retardant and water caused the fire to hesitate and keep burning to the west-northwest, starting to move around the Tuscarora property. The flames were not abating; they were only mildly shifting their course. It appeared to be enough.

What no one could see was a remote outbuilding on the Tusca-rora property. The fire had caught the edges of the small building and within minutes the entire structure was ablaze. It was the first structure to succumb to the flames. It would not be the last. But for now it burned in anonymity.

The rest of the arc around Tuscarora was holding. The wildfire, fed by strong winds, kept rushing to wherever it could find some-thing to burn.

From looking at the maps, listening to the pilots' radio chat-ter, and what they could feel themselves about the wind, everyone thought the wildfire was racing in a west-northwesterly direction. That was excellent news, because if it maintained that direction, in about six miles it was going to run straight into the nearside center of the Cavity Lake burn. That fire burned in a kind of large circle, or oval, stretching from south-southeast to north-northwest. Every-one who saw the maps agreed; if this wildfire maintained its rapid pace and direction, sometime after nightfall it would hit the center of that burned-out area, and there would be no fuel left to burn. It would die out as quickly as it started.

If the wind did not change direction. The signs looked favorable.

• • •

By 12:13 p.m. dispatch had alerted Nancy Koss, Cook County emergency management director, about the start of the fire. Nancy had been the emergency management director since 1993. Among her numerous duties was to make sure Cook County and all its players, public and private, were properly identified in the *Cook County Emergency Operations Plan*.

Even before the 1999 blowdown, she had worked with organizations throughout the county on matters involving fire—including the USFS, the volunteer fire departments, the DNR, the BIA, and many others. She helped coordinate Firewise Communities Program meetings between USFS officials and county property owners. Firewise was a program that taught cabin, home, and business owners how to better prepare their properties so they could survive a wildfire. She also worked with the county's numerous volunteer fire departments to help them get the training and equipment they needed.

While the current conflagration had only been burning for a very short time, if it grew into something serious, numerous officials and departments throughout the county would be affected and could, moreover, be called upon to assist. After being notified, one of the first things Nancy did was begin working down her network of county departments and officials, making sure to contact everyone with a need to know so that they were aware of the potential for a serious incident.

Wildfires like the one beginning to bloom in the north woods bring lots of disparate groups together for different purposes. Because there has been organizational preparation and a long history of working together, on both real and simulated emergencies, Cook County Emergency Management, the GTVFD, the USFS, the DNR, the Cook County sheriff's office, several private contractors, and others have a relatively good idea of who is responsible for what. Under the county plan, the Human Services Department helps feed, house, and care for evacuees and others affected by the incident.

The USFS is the lead firefighter. The GTVFD is there to help fight the fire, but their most important task is to protect structures and help the sheriff's office with evacuations. If the fire burns on DNR land, the DNR takes the lead. If it burns on USFS land, like this fire, the USFS takes the lead. But both help each other in fire suppression efforts. And if necessary, other organizations—the BIA, nearby volunteer firefighting departments, Border Patrol, and so on—have people and equipment they can contribute to the fray.

For simple events, or slow-moving fires, where everyone has a chance to notify the proper authorities and get everyone in place, on time, attending to their respective duties, the chains of command and responsibilities are clear and easy to navigate. But that almost never happens, and it was especially untrue in this fire, which in a little more than an hour had grown into a major conflagration threatening people and structures. And it was only getting worse.

At 12:14 p.m. Engine 573 from the Gunflint station arrived at the scene. Volunteer firefighter John Silliman heard the page from Cook County dispatch and responded by gearing up and driving the engine to the crossroads. Engine 573 is a rig like the one Mike Prom and Don Kufahl were operating. It wouldn't be the last to join the firefight.

At 12:21 Mike Prom radioed Cook County dispatch to tell them "the fire is less than a mile from the Gunflint Trail."

At 12:23 Sheriff Mark Falk notified dispatch and told them to ask Deputy Sheriff Doug Rude to head up the Trail.

At 12:31 a lineman for the local electric cooperative called dispatch, learned more about the fire, and knew he needed to head up the Trail, where he might have to secure the power for any cabins or other structures that became damaged. Dispatch conveyed that information to Mike Prom.

At 12:36 dispatch took a call from Bonnie Schudy at Gunflint Northwoods Outfitters. Bonnie told dispatch about the USFS group of eight on Ham Lake, identifying seven of them by name. (She suspected the USFS firefighters knew at least one of them, which would personalize her call.) She also told dispatch about the pair of canoeists who had gone in just that morning. They were all headed into Ham Lake and beyond. Dispatch told her they would relay the information to the on-site USFS firefighters, who would see if they could locate the campers and determine next steps.

At around 1:00 p.m., Sue Ahrendt decided it was time to get her kids out of harm's way, and sent them with two staffers down the Gunflint Trail to Hungry Jack Outfitters, where they had friends. She told the staffers to return. If they could get back to Tuscarora, she was certain she would need their help.

In the approximately ninety minutes from the first report of the fire, there was a growing response to the emergency, sometimes cacophonous, disjointed, and confused but always following an approach that began to surface an organized community of public and private volunteers and professionals, most of whom knew their roles and responsibilities and all of whom came together for the single-minded purpose of fighting a wildfire that could, potentially, have devastating effects on the forests and people that surrounded this part of the Gunflint Trail.

Because of the wind, the heat, the low relative humidity, and the dry conditions, the fire was burning fast and intense. And it was shockingly early in the fire season to have this kind of blaze this far up the Gunflint Trail. Normally, early season conditions were not conducive to wildfires, so even though the fire season had started and the response had been good, there was understandable chaos as the fight unfolded.

Because of the location of the fire, the air attack was in some areas the only way to fight the flames. At 1:34 Jody Leidholm requested a "quick strike" of CL-415s from Ontario. They were dispatched from Thunder Bay and made a few drops on the fire until they were diverted to another new fire in Ontario.

A couple hours after the second CL-215 arrived from Brainerd, it was flying low, about 150 feet high, to drop more water, when its wing struck an eagle. The sickening thud was unmistakable. The bird was so big that if it had struck the plane at the right place, it could have been as tragic for the pilots as it was for the eagle.

The pilot radioed Jody. "Air Attack One this is Tanker 263. An eagle just hit our wing, but I think we're OK."

Jody considered the situation. If the pilot thought he was OK, it might be reasonable to keep flying. But Jody had been fighting fires from the air long enough to know minor mishaps could cause serious problems down the road. Standard protocol dictated that after a serious plane strike, the plane should be grounded and checked out before continuing to fly. And Jody had lost pilots before, so his decision didn't take long to make.

"Tanker 263, better head back to base and get it checked."

"Will do, Air Attack One."

Jody called dispatch and requested both of the CL-215s out of Bemidji.

The eagle accident was an unfortunate reminder of the chaos that can pervade a wildfire, for humans and wilderness alike.

§9 WITNESSES

At 12:45 p.m. USFS law enforcement officer Barry Huber radioed Cook County dispatch to let them know he was heading up the Trail. Barry was short and fit and wore his black hair close cropped to his head, in the style of a recent armed services recruit. Barry was a USFS veteran; he had been in the USFS as a law enforcement officer for several years and had participated in a variety of investigations. In fact, he was one of the officers the USFS enlisted to train new law enforcement recruits. Today, he had a new recruit, David Spain, with him and was showing him the ropes.

At 11:30 a.m. Barry's USFS radio began crackling with numerous reports of a wildfire. For the first hour, Barry and David were occupied and unable to drive up the Gunflint. But given the weather and what he heard on the radio, he already suspected a person had started this fire.

Across the rugged Canadian Shield, lightning often delivered the first spark of a wildfire. But the day had been dry and windy, and there hadn't been a suggestion of lightning anywhere in the area. At least around the Gunflint, the atmosphere was relatively calm, storm-wise. Fires needed a spark to start. Barry thought the wildfire now being tracked on his radio was most likely camper ignited.

•   •   •

By the time Barry and his trainee arrived at the Tuscarora turnoff, numerous official vehicles and people crowded both sides of the Gunflint and the start of Round Lake Road.

Once Barry saw the crowd, he knew his presence would only add to the congestion. So he moved his vehicle back down the Gunflint Trail, where he could monitor the activity and respond when the proper occasion arose.

For the next hour or so, Bonnie Schudy from Gunflint Northwoods Outfitters fretted about the two parties she had outfitted. She also worried that in all the commotion, her message had not reached the people who needed to check on the safety of her folks. Her worry that her request would get lost in the mayhem finally compelled her to drive up to the turnoff, where around two o'clock she encountered the growing throng of civilians and firefighters, including Sheriff Mark Falk, who had recently arrived on the scene. When she asked Sheriff Falk if he had received the message from Cook County dispatch, he told her he had not.

When Bonnie noticed all the USFS personnel on-site, she conveyed her concern to them. She knew some of them would probably know their colleague Trent Wickman, one of the USFS Air Resource Management people who was with the party of eight on Ham Lake. Trent's office was in Duluth, so he was no stranger to the area.

Vance Hazelton was the on-site IC. Vance turned to Dave Snyder and Jeremy Rux, two of his firefighters, and told them to backtrack down the Gunflint Trail to Kings Road, where they could get close enough to Ham Lake to bushwhack into the eastern side of the lake, behind the fire. Unfortunately, neither of them were familiar with the back trails that would take them into the eastern side of Ham Lake.

Mike Prom remembered Sue Ahrendt had already tried to enter Ham Lake via BWCAW entry point 50, where she wanted to check on their lone camper, Steve Posniak, but was turned back by the flames. Now Mike turned to John Silliman, whose day job was head naturalist at Gunflint Lodge. John knew Kings Road and the other trails that existed in the area. If you knew where to look for the trails and which way to walk once you found them, you could bush-whack along old trails through the woods all the way to Ham Lake. John agreed to accompany the two USFS fighters and show them the way.

"Just make sure everybody's safe," Vance said. He told them to keep him abreast of progress via their radio.

Snyder, Rux, and Silliman took off down the road with a canoe, heading for the turnoff. Within fifteen minutes they were in the early spring woods, carrying a canoe south, searching for the first glint of water through the trees.

On Ham Lake the USFS group of eight, who had left their campsite around ten o'clock, paddled four canoes across the water to the short portage into Cross Bay Lake. The north end of Cross Bay Lake is long, twisting, and narrow. The first section slants in a southwest-erly direction for almost a kilometer. Even though this north arm of Cross Bay Lake is narrow, with the forest coming up close enough on both sides to act as a windbreak, it was still tough paddling. After a sharp turn east, the narrow waterway continues almost due south for about half a kilometer. At this point the journey into the rest of Cross Bay Lake continues by making a sharp turn to the southwest. However, by continuing in a straight course, paddlers often got off course by entering Extortion Creek, a waterway that looks like the continuation of the one they had been paddling. On a map, the junc-ture appears obvious. But unless you have a compass and are fol-lowing a map, it is easy to be seduced by the beauty of water, trees, rocks, marsh, and wildlife.

This day, the flotilla of four canoes continued paddling into Ex-tortion Creek, following the waterway until it began to narrow and they realized they were in the creek. As they all paused to consider their location, the two Ohio canoeists paddled up behind them. The USFS group had noticed them earlier; they had passed by Steve Posniak's campsite earlier that morning. They, too, had become lost and now wondered if they had missed an important turn.

After a brief conversation, the USFS group told them there was no portage ahead and if they wanted to get to Cross Bay Lake, they needed to turn around and take the waterway they had missed. Trent Wickman also let them know they were all now in violation of BWCAW regulations—with five canoes and ten people they exceeded the four canoe and nine-person group maximum. He told them to go ahead. The two Ohio paddlers thanked them and turned around, paddling back the way they had come.

Here the waters were calm. Trent's wife, Kim Wickman, decided it was a good time to snap a photo. She focused on the man-woman team in one of the other canoes and clicked. She knew the time stamp on her camera was off by fifty-three minutes, which meant the actual time was 11:16 in the morning.

By now they had checked their maps and seen where they had made their mistake. Once the Ohio canoeists were gone, they, too, decided to turn around and paddle back the way they had come.

Two USFS women were in the lead canoe. Ahead they had an unobstructed view up the channel, and they stopped, motioning for the others to come up beside them. When the others caught up, they pointed out the large smoke plume coming from the vicinity of their campsite.

Kim Wickman figured at least fifteen minutes had passed since the two Ohio canoeists had paddled away, which made the current time around 11:30.

The group decided to return and make sure their gear was safe. They turned, paddled hard, and covered the distance to the short portage into Ham in less than thirty minutes. By the time they entered Ham, they could see the wind direction was blowing away from their site, and the smoke plume was coming from around their neighbor's campsite, farther down the shore. Around noon they beached their canoes.

One of their first actions was to head out onto the stretch of rock overlooking Ham Lake. From there, they had an unobstructed view of the westernmost campsite down the shore. They could see the campsite was on fire. Two of the group contemplated taking some cooking buckets down to the site to fight the fire. But while they discussed it, a tree flared and crowned, and they realized they did not have the resources to fight a fire that was already that intense.

They could not see the camper or his canoe or supplies, but they could see trees torching and extreme fire behavior.

At this point they were all anxious to alert the authorities to the fire. It was windy and dry—terrible conditions for a runaway wildfire. One of them suggested he and another could take a canoe and try to paddle back over the portages they had taken to get here. But that direction—heading toward entry point 50—would take them straight into the fire and smoke. From what they could see, it would be foolish to paddle into that dangerous zone. So for now their thoughts of alerting the authorities were frustrated.

Approximately a half hour after they returned to their site, they heard an overhead drone and looked up to see the first plane come onto the scene, a Beaver. This was excellent news, since the Beaver was obviously on-site because of the fire and smoke. The authorities must now be aware of the growing conflagration, so the USFS group decided that for now, all they could do was break camp, pack up, and be ready to move, in case the wind direction shifted. They set about the task and got ready to enter their canoes with their equipment, should the need arise. While they were packing their gear and eating lunch, they continued to watch the rapid pace of the fire and its smoke plume.

Not long after the Beaver plane arrived, it was followed by the fixed-wing aircraft with Jody Leidholm and a CL-215. Later, more planes arrived and took up the fight.

By midafternoon, the fire front had moved well downwind, and the plume dwarfed or concealed all the aircraft flying northwest of the USFS group.

•   •   •

On the north side of Ham Lake there were four campsites. Steve Posniak's site was the farthest west. A quarter mile down the shoreline was the site now occupied by the USFS group, near the middle of Ham Lake. To the east of that site there was another, and beyond it was the site farthest east on the lake.

John Silliman led his group down Kings Road, then onto the Ham Lake ski trail. The ski trail led to another offshoot trail, which John and his two companions took, bringing them to the lake. At around three o'clock John and USFS firefighters Dave Snyder and Jeremy Rux put in at the easternmost campsite.

John had been on Ham Lake many times. One of the first things he recognized, putting in on the east side of the lake, was the wind. It was blowing hard and fast and created foot-high whitecaps on the water. In fact, not once during all of John's other traverses of Ham Lake could he recall seeing such big waves. He was thankful he was with experienced canoeists.

They carefully paddled down the lake and found the USFS group in the central campsite. They beached their canoe long enough to discuss next steps. Jeremy radioed to Vance Hazelton to let them know they had found Bonnie Schudy's group of eight, but that there appeared to be no sign of the overweight camper at the westernmost campsite. Vance told them to evacuate the group the way they had entered, via the trails through the woods out to the road. He told them to provide the group with whatever support they needed.

While Jeremy helped the campers gather their equipment, John and Dave paddled to Steve Posniak's campsite. What they found surprised them.

First, it appeared certain this was the site from which the fire had originated, but it was also clear that it did not start from the fire grate. The area around the grate was not burned, there were still paper and other unburned debris around the edges of the fire pit, and the remaining ashes and coals were wet and cold. There were two or three places on the site still burning, but mostly the fire had spread to the west-northwest, across the narrow river channel, where they could see it was continuing to make rapid progress, leaving behind smoldering woods and a rising smoke cloud. They spent ten minutes examining the site, doing some fire suppression, and stringing up orange tape, indicating the site had been checked and was now closed. Then they returned to their canoe and paddled back down the shoreline to the nearby campsite.

By the time the two returned to the USFS group's site, everyone was ready to evacuate. They all pitched in to gather their canoes and gear, and John led them down the shoreline, paddling to the easternmost campsite. Once they put in at the site, they hauled their equipment and canoes out of the water and started hiking out of the woods. It took thirty to forty-five minutes to slog their gear to Kings Road, and they did it in one hard haul.

While the group was still on Ham Lake, Bonnie Schudy found out approximately where and when they would be coming out, and she and her assistant, Brian Gallagher, drove two trucks down Kings Road to meet them. Not long after the group reached the road, Bonnie and her assistant arrived in the trucks.

While they were loading up their gear and being transported back to Gunflint Lodge, members of the group told Bonnie about the Ohio canoeists, and how they had all taken the wrong turn on Cross Bay Lake. They also told her they thought the Ohio group paddled back to the turn into Cross Bay and headed south. Given their location, they would be paddling and camping in almost the opposite direction from which the fire was growing and spreading, so Bonnie felt comforted that for now at least, her Ohio group was out of harm's way.

The group also told her about the lone camper. None of them knew his name, but they had all seen him and could describe him. They also mentioned that two in their group had spoken with him just yesterday afternoon, when they first entered the open water of Ham Lake.

•   •   •

By around four o'clock, Dave Snyder, Jeremy Rux, and John Silliman were back at the Gunflint Trail, on the turnoff for Round Lake Road. Here they met USFS law enforcement officer Barry Huber. Dave told Barry that the USFS group they had evacuated had information about the start of the fire. Dave also told Barry that he had visited Posniak's campsite and that it appeared to be the place from which the fire started. He told Barry what he had found and that he had marked it off with orange tape.

When Barry learned that the USFS group had information about the fire's origins, he knew he had to speak with them. And he expected he eventually would have to paddle into the campsite and have a look for himself. But he suspected he already knew what he would find. That made the overweight camper on Ham Lake a suspect. If he had a suspect, Officer Huber would need to get others involved, particularly from a fire investigation perspective.

He decided to head to Gunflint Lodge and find out more about what the USFS group of campers had seen and heard. So far he had heard nothing about the whereabouts of the lone camper. He would begin with the next best thing, the nearest he could find to witnesses of the start of this growing conflagration.

10

§ AN INTERRUPTED JOURNEY

On Eagle's Nest Island, Lee, Layne, and Gus could feel the increasing fury of the wind. On top of the rock, the three had to lean into the wind to stay vertical. At one point Lee and Gus leaned out over the gradual fall of rock, and the wind was strong enough to support them at an unnatural angle. Layne moved off to the side and took a few photos of his companions leaning at a preternatural angle out over the cliff edge.

Then, off to the southeast, they heard the drone of a plane coming from the north, an unusual occurrence since flight over the Boundary Waters is banned. They watched a yellow CL-415 come from Canada, heading straight for the fire.

"They're already getting air support," Lee said.

They watched the big yellow plane disappear into the southeast. The Canadair CL-415 SuperScooper was an unwieldy-looking aircraft with a fat underbelly designed to drop into water, scoop up more than 1,500 gallons, and drop it on wildfires. The three listened to the plane's drone over the wind, heading straight for the column of smoke.

When they finally took a moment to gaze behind them, over the large expanse of Seagull Lake, where they were planning to paddle, they could see whitecaps on the water.

"I don't think we can make it over that," Gus said.

"I think it would be foolish," Lee agreed.

"I'd never try it in a canoe," Layne said. It looked like it could be fun in a kayak, but not when the water was this cold.

The three considered their options. They could see the smoke of the fire growing, and from what they could tell, it appeared they were in its path. The wind was still strong and gave no indications it was going to abate anytime soon. In fact, the day was growing warmer; the unusual early spring heat would only increase the atmospheric turbulence.

"Let's try and make it to the palisades," Layne said, peering to the northeast.

Both Layne and Lee had camped near the palisades. On the map the oddly shaped Seagull Lake peninsula jutted down into a point that curled at the end. The shape and direction of the point angled to the southwest, forming a high-rise wall. The designated campsite was tucked into the nook of the peninsula. The position would provide them with excellent protection against the southeasterly squall. And from there they could climb up to the top and have a clear vantage point from which to watch the fire. They were hoping the fire wouldn't burn the peninsula, but if it did, they could paddle to Miles Island, and from Miles farther west into the wilderness bordering the northwest side of Seagull.

The only problem: the crossing did not look easy. A string of islands stretched away from Eagle's Nest to the northeast. For parts of the passage they could get in behind those islands and be partially protected. But there was a lot of open water, too, with whitecaps cresting the lake's surface. They would have to be very careful, and Lee and Gus weren't looking forward to the passage. There appeared to be at least a mile of open water they would need to paddle through, much of it angry. Gus, the more experienced paddler, could feel his anxiety rise at the prospect.

"If we're going to do it, we'd better get started," Gus said.

They descended to their canoe and kayak and in the protected inlet got situated in their respective crafts. The wind was still high, and when they gazed out at the water they were about to enter, it appeared worse than when they had considered it from atop Eagle's Nest. In the distance they knew the palisades were protected, and it was the best place to hunker down in front of an oncoming fire. There was a smoke wall building to the south. The burgeoning fire on the horizon was growing so quickly they thought it was only four to five miles distant. In reality, it was nearly ten miles to the southeast. And the truth was it was headed in a westerly direction, blown by winds out of the east-southeast. But from appearances, it looked as though they were likely in its path . . . or would be. And sooner rather than later.

Wildfires move at speeds that are dependent on physical factors like wind, combustible fuel, temperatures, and precipitation. This one appeared to be coming on fast. But again, the massive smoke wall to the south where just this morning there had been clear sky made it appear nearer and moving more quickly than it actually was. The fire would take a while to cross the parched forest to the west before it hit the burned-over country of Cavity Lake— that is, if the wind didn't shift and bend the fire's trajectory in a different direction.

At any other time, Layne, the seasoned kayak paddler, would have enjoyed the challenge of the crossing. But now he worried about his companions in their canoe. After crossing a narrow opening, during which both crafts were buffeted by the wind and waves, they came in behind a short island and slightly more navigable seas. But only for a moment. From here they could see there was a relatively long stretch of open water before the next larger island, and then beyond it a still longer expanse of whitecaps.

When they pushed into the first unprotected water, they were immediately hammered by high wind and waves. And the water was cold. Gus knew that if they capsized, their vests would keep them afloat but doubtfully long enough to reach a nearby shore and safety.

Their vests would serve only to keep their lifeless bodies afloat, so they could be recovered after they finally succumbed to the cold.

In the open water it required all of Gus's canoe skills to keep the craft upright. In the bow, Lee was struggling. There was an echo effect when the large waves bent around the islands, causing swells to buffet them from both sides. They would paddle, paddle, and then be struck by a wave that threatened to flip the canoe. Water poured over the bow from both sides, dousing Lee. The canoe flipped right, buffeted by a blast. Then left, buffeted by another echo wave. And then they would paddle like hell before the next swell broke.

Gradually, painstakingly, feeling something akin to terror, they made headway toward the next small island and slightly calmer waters. The wind was still very strong and showed no signs of abating, its fury whipping the water into a frenzy. And in the distance, almost ten miles away, it was having a similar effect on the fire, feeding the flames with rapid drafts of oxygen that when combined with the dry forest sent columns of thick smoke billowing into the sky and a growing fire rushing forward to feed upon the next parcel of parched tinder.

After passing the relatively navigable waves of the next island, they entered the largest expanse of open water. With every meter the wind and seas threatened to capsize them. They both understood there was nothing to be done. At one point Lee worried they were going to take on water that would hopelessly maroon and capsize the small craft.

"Paddle as hard as you can," Lee yelled back to Gus. "There's nothing we can do now. We can't turn around. There's nothing we can do but hope we make it!"

Fear forced them to focus on their struggle through the waves. The second-by-second paddle, driving into the water, pulling forward, watching the waves, fighting the wind, occupied them with an intensity that allowed few other thoughts.

Finally, after a struggle neither of them wanted to relive, they reached the calmer waters behind the curl of the palisades.

Much later, in retrospect, Gus would recall the crossing as "one of the most frightening moments of my life . . . pretty insane. Thank god for Ted—that's a really stable craft." Ted Bell was the renowned builder of Gus's canoe.

When they arrived at the palisades, they felt tired and thankful. Overhead, the height and curve of the peninsula acted like a big wall, cutting off the wind. The campsite was down by the water, largely sheltered from the intense breeze. It was a good place to ride out a wildfire, if they ended up in the trajectory of its oncoming path. And it was a five-minute climb up a trail to the top of the cliff, where they would be able to keep an eye on the approaching smoke and flames.

Now they needed to set up camp. They were safe, for the moment. The only thing that would drive them back into the waves would be oncoming flames. If that happened, they would take their chances in the waves. But when they looked out at the water, the whitecaps appeared menacing. They hoped they would be able to stay here, and they hoped the fire—if it came—would burn over or around them. They would just have to wait and see.

When Gus peered out at the churning waters, he considered a paddle into it near suicide. And he let the others know he wouldn't be leaving unless forced. "I have a wife and kids, and I want to see them again. I don't want to die in the middle of Seagull Lake."

His companions agreed.

Since morning, the rising smoke ten miles to the southeast had grown from a mistaken campfire to signs of a wildfire to something much bigger: a smoke wall building into the stratosphere, being pushed up overhead. The air gradually fogged over with light smoke. While they retrieved their gear, the first ashes began to fall. They were puffy and light gray, like big snowflakes that drift down from a winter sky. The smoke and ashes were increasingly punctuated by more overhead planes, either flying to the front lines to drop water or retardant, or returning to replenish their loads.

As the afternoon progressed, so did the air traffic. The acrid smoke fog periodically settled over the camp, burning their throats and eyes.

11 FRONT LINES

USFS Beaver pilot Dean Lee brought the plane in low over the big blue of Devil Track Lake. There is a small airport on the north side of the lake. The seaplane dock is farther down the road to the west. On this day the winds were strong out of the east. There were whitecaps on the water, but it was nothing, Dean knew, the Beaver couldn't handle. He brought the plane down carefully and settled onto the bumpy surface, taxiing to the dock where Kurt Schierenbeck was waiting. It was 1:35 p.m.

As a Type III IC with significant firefighting and organizational experience and expertise, Kurt knew he needed to get above the fire where he could obtain a more comprehensive view. He wanted to see the fire's intensity and the types of available fuel, and determine how much land had already been burned. Importantly, he needed to figure out—weighing wind direction and its burn path—where the fire was heading. And he needed to determine what roads and trails were in the vicinity that could be used to shuttle firefighters and supplies. As a longtime resident of the area, he knew the region, but he wanted to make sure he had a clear idea of everything from footpaths to ski trails to USFS roads to highways, all of which became suddenly important as ways the fire could be surveyed as well as fought. And there were cabins and resorts in the area. Kurt needed to see what other structures might be threatened.

As soon as Kurt was airborne, the chatter began. He had a radio on his belt, a cell phone, and while in the plane, radio headphones. His headphones were busy with multiple conversations, and he responded to whatever he could as he and the pilot rose into the afternoon air.

Near Devil Track, it was clear, sunny, and windy. But Kurt was familiar with how dramatically weather could change when moving up and away from the big water of Lake Superior to the interior of Minnesota's north woods. There had been times during the winter when he had left his house down by Superior to go ice fishing on a lake fifty miles inland. At his house the temperature might be ten degrees, while fifty miles inland it was thirty-five degrees below zero. The difference in temperature was just one of the reminders that the weather in Grand Marais may not translate to the weather around Ham Lake. You must be prepared.

Devil Track Lake is approximately six miles inland but much higher in elevation than Lake Superior. As the pilot and Kurt rose in the air, climbing in elevation and turning the plane in the direction of Ham Lake, it didn't take long before they saw the fire's smoke plume, far off in the distance.

Over the next fifteen minutes the smoke plume grew from a distant white stretch to something much more substantial. Once they got near enough to see fire details, Kurt's alarm grew. The wind was intense, and the running crown fire was burning fast and furious. He was slightly comforted by the fire's direction—nearly due west—into the Cavity Lake burn. But he also noticed the fire was spotting.

While he was in the air, circling the fire, getting a better look and beginning to determine the fire's size, his radio continued to chatter. One of the chatter threads was Vance Hazelton, down on the ground at the turnoff to Tuscarora.

There was a lull in the planes dropping water and retardant. Vance saw the fire had made its pulse across the road. The main front appeared to have moved beyond the main structures around the lodge.

"Is it safe to go back to Tuscarora?" Vance asked. "We need to check on the place and get those sprinklers running."

It was so early in the season and the fire had come up so fast, Andy and Sue Ahrendt hadn't had a chance to start their sprinkler system, though it wouldn't take much to get it operational.

The smoke covering the area made it hard to determine if any of the Tuscarora buildings were on fire. Kurt caught glimpses of the rooftops. He could hear the chaos on the ground. Tuscarora's structures were the fire's first protection points. So far the retardant and water had done a good job bending the approach of the fire. But they needed to do whatever they could to further protect the resort. Finally, at around two o'clock Kurt got a good enough view of the area to give Vance the go-ahead to enter.

Before driving back up the road, Vance and the others considered exit strategies, just in case they drove up the road and flames closed in behind them. For now, all of Tuscarora's canoes and watercraft were still up at the site. If the unexpected happened and they needed to make a run for it, they could always get into a canoe and head out onto Round Lake.

In moments, Vance, Pete Lindgren, and Tom Lynch drove up the road to check on the grounds, put out spot fires, and get the sprinklers operational.

By now there were even more vehicles at the corner of Round Lake Road and the Gunflint Trail. Acting GTVFD chief Mike Prom and Sheriff Mark Falk remained at the corner, directing traffic and fielding calls.

While the crew headed into Tuscarora, Kurt was bumping around overhead. On his lap he had a notebook in which he was sketching the geometric shape of the fire, and how much land had already been burned. And from this height he was also noticing *how* the fire burned.

For starters, the shape of the fire was like a big cigar, very narrow at one end, where the fire had begun, and at the other, starting to widen. And the flames were spotting. He could see from the air that the wind was blowing the fire in a way that picked up embers and carried them to some distant place in front of the main burn, where they found purchase in the dry forest, igniting a new burn.

Doing a series of calculations, he determined that since 11:30 a.m., the fire had burned 480 acres. He relayed the information to dispatch in Grand Rapids, so they had a sense of the rapid and growing rate of the fire. He also let them know it was a running crown fire and spotting.

Kurt was in the throes of the action now. He was bouncing in the air, figuring out the fire size, making observations, and all the while fifteen people were chattering in his ear. On the ground he had three excellent colleagues—Vance Hazelton, Tom Lynch, and Pete Lindgren—to whom he had delegated the responsibility for managing the engines, personnel, dozers, structure protection, fire scouting, and more. It helped to have his capable colleagues on the ground, but the fire's behavior was so aggressive, it was still chaotic.

At one point he looked down and noticed the sparks flying and saw a fifty- to sixty-year-old stand of red pines. The fire was traveling so fast and spotting that it had burned all the pine in the center of the timber stand. The burned timber was black. The surrounding timber was green. It was odd behavior for a wildfire but indicative of how hard wind, flames, and tinder-dry fuel could combine to produce unexpected, inflammatory results.

Part of the reason Kurt was keeping USFS dispatch in Grand Rapids informed was because of protocol; it was standard operating procedure. But another reason was because since first setting eyes on this blaze, he had suspected it might be a wildfire that required a higher, more robust response than a Type III incident. If his over-sight continued to feed his growing perspective, he would need to talk it over with the district ranger.

USFS protocol required a district ranger's OK and signoff before elevating a fire from a Type III to a Type II. To make that kind of call, the district ranger required solid background information about how quickly the fire had started and spread, and what threats it posed to structures and the rest of the forest in front of the fire's path.

In addition to Kurt's years of practical firefighting experience, he had also spent many hours in the classroom. He knew, for example, the protocol for elevating a fire from a Type V to a Type IV, Type III, and higher. Obviously, the characteristics of wildfires needed to be much more complex and threatening the higher the elevation on the IC Type chain. The details and characteristics for each fire type were spelled out in the U.S. Department of the Interior's Redbook (*Interagency Standards for Fire and Fire Aviation Operations*). Updated annually, the Redbook was the bible for managing the idiosyncratic chaos of a wildfire.

Kurt was familiar with the IC Type characteristics enumerated within its pages. Chapter 11 of the Redbook was "Incident Manage-ment and Response," which included subsections about "Wildfire Risk and Complexity Assessment" and "Incident Characteristics." The chapter also referenced Appendix E, which contained a Wild-fire Risk and Complexity Assessment form, and Appendix F, which described indicators of incident complexity.

Due west, the fire would be heading into forest, so the immedi-ate threat to structures was confined to Tuscarora. But beyond that forest there were more structures that could be threatened. And if the wind made a subtle shift north, it could crawl around the Cavity Lake burn and begin pushing the fire in a more dangerous direction. If the fire moved farther north, the potential for damage to struc-tures and the threats to people living up the Trail could be signifi-cant. Kurt wanted to make sure the USFS took every precaution to keep people and property safe.

The USFS fire suppression philosophy was also clear. If fires are started naturally, they are not immediately suppressed but managed for "resource benefit." However, if the fire was human caused, suppression efforts were almost always immediate and comprehensive. At this point, no authorities knew how the fire started, but because of the weather, Kurt was almost certain it was human caused. No one had stopped to pause in the suppression efforts, because structures were clearly threatened. Even if the fire had started naturally, the USFS would do everything in its power to protect the public and property.

If the fire was human caused, which seemed extremely likely, then the USFS would go all out to suppress it.

Back at Tuscarora, Vance got a call from Kurt in the air, requesting that he meet him at the Seagull Lake dock. He needed to meet with Vance to get a feel for the fire on the ground and to get a ride to the nearby Seagull Guard Station. If Kurt was right and they needed to elevate this to a Type II IC, they could begin making calls from the station.

Within a half hour of the call, Kurt had landed, and Vance was waiting for him. They raced up the road to Seagull station, discussing what Vance had seen on the ground and what Kurt had seen in the air. By the time they reached the station, Kurt had realized this fire was already extreme enough to require more equipment, supplies, personnel, and logistics than he was able to manage. He called the district ranger for the Gunflint District, the most senior person at his office. Barely three hours into his fight of the Ham Lake fire, Kurt requested that the response be elevated from a Type III to a Type II.

Dennis Neitzke was the permanent district ranger for the Gunflint District. Unfortunately, Dennis was on temporary assignment at the Chequamegon-Nicolet National Forest in western Wisconsin. The Gunflint District's interim district ranger was Carlene Youkum, who did not have the firefighting experience or expertise required to elevate the response to the fire from a Type III to a Type II. Nevertheless, Kurt convinced her to work with her colleagues to get the designation elevated so the Type II team could be called up. She did that, and since for now she was the nearest district ranger to the fire, and since she could at least communicate what she knew about the fire if she was able to see it, she decided to get a better look at this fire. Kurt let her know he would pick her up at Devil Track Lake in the Beaver as soon as they could get down there.

Now that Sheriff Mark Falk was on the scene, he was facing increased pressure to call for more evacuations. But he, too, felt like he needed more information before making such a serious and important decision. When he found out Kurt was heading up in the Beaver, he decided to accompany him, to get a bird's-eye view of the fire.

Over the next hour Kurt, Mark, and Carlene viewed the fire from above. It showed no signs of abating and had consumed much more forest since Kurt's first estimate of 480 acres.

The Beaver set down on Seagull Lake and dropped off the sheriff. Then they were up again and heading back to Devil Track Lake to drop off Carlene and Kurt. Kurt's truck was at Devil Track, and he decided he would need it if he was going to manage the fire from the ground, at least until the Type II team arrived.

USFS district rangers report to a forest supervisor. At this time, the forest supervisor for Minnesota was Jim Sanders. Since Carlene was technically not qualified to determine if the response to this fire should be elevated to a Type II, Forest Supervisor Sanders found the nearest district ranger with the requisite experience to make the determination. In this instance the most logical person was John Wytanis, the district ranger from the nearby Tofte Ranger District.

By the time Carlene returned to the Gunflint District headquarters, she had communicated what she had seen to John. Since there was much to learn, Carlene would shadow John during his deliberations. Earlier, Kurt's request for a Type II team had been granted, and the Type II team had already been called up.

The Type II Incident Management Team was comprised of approximately twenty interagency firefighting experts from all over Minnesota. Members of the core team were responsible for command, finance, logistics, operations, and planning. And that was just the core team; many currently fighting the fire would stay on, and more would be added, as the Type II team began to get a feel for the wildfire they had decided to take on.

The lead Type II Incident Commander was John Stegmeir. In Minnesota there are three Type II teams. During fire season, each team is on call one week and off for two. The team that responds when a Type II team is needed is the team that's on call. Coincidentally, John's team had been on call for the Alpine Lake fire, the Cavity Lake fire, and the East Zone Complex fire. Consequently, John and his team were familiar with many of the players already working on the Ham Lake fire.

John worked for the DNR Division of Forestry out of the Orr, Minnesota, office. When the call came in for a Type II team, John was on fire watch in the Orr office. Regardless, all the on-call Type II team members, who worked for different organizations all over the state, had their bags packed and ready so they could leave at a moment's notice. They tossed their bags into their vehicles and headed to Grand Marais. John, like his colleagues, threw his bags into his Explorer and started down the highway.

While the Type II team had been called up, Kurt was racing back up the Gunflint Trail. It would be much later in the evening before most of the Type II team members arrived. Kurt would need to brief them about the firefighting efforts (and the fire's behavior) to date. So for now, he was still the incident commander for this fire, which meant he had to get up to the front lines and continue monitoring the fight.

Around 4:30, Andy Ahrendt and then Sue Ahrendt were both escorted to Tuscarora, where they immediately began working with Pete Lindgren, Don Kufahl, and Tom Lynch on fire suppression, setting up dozer lines, working on sprinklers, and more. Sue worked with her two staffers to begin loading their Kevlar canoes onto trailers. Kevlar is basically plastic. If fire touched their Kevlar canoes, they would burn hot and fast with noxious smoke. Once they had filled their trailers, Sue and her two staffers headed to Gunflint Lodge, where she was hoping she could park them until the fire was entirely subdued or had moved well beyond the Tuscarora grounds.

12 QUESTIONS

After the USFS party returned to the Gunflint Lodge from Ham Lake, Bonnie Schudy, head of Gunflint Northwoods Outfitters, checked them into the Gunflint Lodge's canoe cabins.

While the group was getting settled, Sue Ahrendt stopped by and asked if they could park Tuscarora's Kevlar canoes on the Gunflint Lodge's grounds, to keep them safe.

Resort owners along the Gunflint share a communal sense of life in rugged surroundings and the knowledge that on occasion circumstances beyond anyone's control can go awry. It's good to give a helping hand, not simply out of a sense of altruism but also because in environs like these—when wolves are literally howling at the back door—there is no telling when one's own businesses or lives may be threatened. They are definitely competitive for the travel and outdoor adventurer's dollar, but they are also neighborly.

Bonnie not only let Sue know where to stow her Kevlar canoes, but also asked if there was anything she could do to help out. Bonnie was familiar with Tuscarora Lodge because earlier in her career she used to work there.

The truth is, there was more urgent work than hands at Tuscarora. Spot fires were blazing up on Tuscarora's grounds, and pine needles and other forest debris needed to be swept and raked away from Tuscarora's buildings, where they could easily spark and lead to a structure blaze. Embers needed to be put out, perishables needed to be removed from refrigerators without power, and more. Bonnie let Sue know she would gather together a crew of five and head to Tuscarora shortly.

Around 4:30 USFS law enforcement officers Barry Huber and David Spain arrived at the Gunflint Lodge. Bonnie's husband was working at the main lodge and notified Bonnie that she needed to stay put, that two law enforcement officers would like a word with her. Bonnie remained at her desk long enough to let Barry know where he could find the USFS group, and then gathered her crew and headed to Tuscarora.

When Barry found the group, he questioned them about where they were and what they saw. According to Barry's Supplemental Incident Report (USDA Forest Service Supplemental Incident Report Number 7503337), "The group of eight said that they had seen a man camping overnight on a site northwest of their location and that he was at that campsite when they first saw smoke early that day. They reported the smoke had dissipated but then around 11 AM, a large amount of smoke was seen coming from the area of the site in question. When asked to describe the man, three people said they had actually talked to him on Friday and they had the best description. Trent Wickman, Ann Acheson, and Ann Mebane described a white male, 50 to 60 years of age, grey hair, overweight, wearing a red flannel shirt and with an aluminum canoe."

In fact, the conversation with Steve was brief, but memorable enough so several from the group recalled Steve's aluminum canoe, relative age, stature, and other details. Part of what both Ann Acheson and Ann Mebane remembered was that Steve was wearing headphones when they paddled past his site. In part, Ann Mebane's written statement said, "At the top of the rock was a white male with gray/white hair facing towards Ham Lake. I'd guess he was between 60 and 70 years of age, probably around 6 feet or so. . . . He was somewhat heavyset (maybe 220 lbs?) and not particularly fit in appearance. I don't think he saw us at that point. He was wearing headphones—I didn't pay much attention to the rest of what he was wearing as I was fixated on why someone would be wearing headphones in the wilderness. . . . I mentioned this to Ann and pointed out the headphones. I also wondered (aloud, I believe) about whether he was alone or not as I could see no evidence of another person. I was thinking about how heavy that aluminum canoe would have been to portage in by himself."

When the two women paddled around the point, they were blasted by the wind. As already noted, upon their return to calmer waters to get repositioned for another run, they briefly spoke with Steve.

The point is, all eight of the USFS party witnessed Steve's campsite, the overturned aluminum canoe, various clothing items hanging out to dry, and especially Steve—describing him in some detail, including his age and weight. And at least two of them commented on the headphones, something Steve enjoyed while in the woods and elsewhere.

Steve did not know these people, where they came from, or who outfitted them. He also did not know seven of the eight were USFS employees. He also could not imagine that in approximately twenty-four hours, fellow USFS employees (firefighters Snyder and Rux) would be sent in with John Silliman to evacuate the campers, and that two USFS law enforcement officers would already be on the scene and would begin questioning them on what they had seen.

Since on the day of the fire the USFS group on Ham Lake had not seen Steve leave his campsite, there was confusion about where he spent the afternoon. The USFS group returned to their campsite in the middle of Ham Lake, with excellent vistas down the lake to the west, around noon. Since Steve had been frustrated by his inability to put out the fire and keep it from spreading, he could have turned upwind to escape the flames. But if so, he would have run into the USFS group returning to their campsite.

He could have quickly paddled across from the point due south and entered a small creek, but again, if he had done so, he would have been in plain sight of the USFS group. Also, it was very windy and rough on Ham Lake, making it dangerous to paddle sideways against the breeze, especially for an infrequent paddler like Steve.

While the two portages along the Cross River were either burned or burning, the river widened directly away from Steve's campsite and was partially protected from an eastern breeze. In fact, the previous day Trent Wickman had made note of its relative calm, describing their trip into Ham Lake when they entered this wide water: "After we got across the second portage, the bay was very calm (as opposed to the trip so far which was windy) so we soaked up the sun for a while and took pictures."

Consequently, the logical place to go—after he had gathered his supplies—was back in some recess along the widened stretch of the Cross River, or what Trent Wickman described as "the bay." Regardless, no one on the ground in the area saw him leave his campsite. And only one person in a plane—Jody Leidholm—thought they saw Steve in a canoe, very near his campsite. But that was only briefly. And all of the planes and the helicopter were flying low enough—descending to drop their loads of water onto the fire and then periodically dropping down onto the lake's surface to refill their tanks—to have easily spotted a canoe.

Steve must have packed up and left his campsite slightly before noon, since that was the approximate time the USFS group returned to their campsite.

Sue Ahrendt had not spoken to any of the USFS group, so she did not know where they had camped or what they had seen, or that Steve had camped at the Ham Lake location. When Steve was completing his BWCAW permit, he had indicated he would be staying on Cross Bay Lake.

• • •

At Tuscarora in the late afternoon, Sue and her team from the Gunflint Lodge and others were raking pine needles away from buildings, loading more canoes, emptying freezers, and attending to other duties. Sometime between 5:00–5:30, Sue saw Steve Posniak walking up the road to Tuscarora from the Cross Bay Lake put-in site. According to her written statement, "He asked for his keys so that he could go back to the Cross Bay landing to pick up his gear. I told him that I was glad that he was safe. I asked about his experience. He told me that the last portage was really charred, and that it had been a dicey morning."

Sue or one of their employees let Andy know Steve had returned. Andy met Steve at Tuscarora's trading post building. When Andy heard Steve's gear was back at the landing, he asked Brian Gallagher, who was also helping out at Tuscarora, to use one of Tuscarora's vehicles to pick up Steve's gear and canoe.

Steve walked back to the landing.

When Brian arrived at the landing, he introduced himself and greeted Steve, helping him with his gear. Brian described the meeting in his statement: "I told him who I was and he began to tell me he had been camping on Cross Bay Lake. I thought it was odd that he tells me this without asking. I began to load his equipment and during the questions about which packs were his he repeats the statement that he was camped on Cross Bay Lake, again without me asking."

Steve returned with his gear. Sue helped him stow it. Her statement in part reads, "Upon return I helped Steve Posniak move his packs to the picnic shelter. He told me that he had been camped for two nights on Cross Bay Lake, that he had come to the Cross Bay Lake/Ham portage at around 8am and had seen the flames. He said he did not see any people."

Andy was also helping move Steve's packs and get him settled. When Andy asked Steve for some additional details, Steve continued his prevarication. In part, Andy's written statement described the exchange: "I asked him where he had camped? He said 'Cross

Bay.' I asked 'Both nights?' He said 'Yes.' I recall we then moved his packs to the picnic shelter and he moved his vehicle adjacent to the picnic shelter where he could unpack his gear. He then pointed to the northwest campsite on Cross Bay Lake on his map and said this is where he had camped. I asked him when he had left the site in the morning and I recall he said around 8:30 a.m. I then asked about his encounter with the fire and I recall he said he had seen the fire on the end of Ham Lake when he was at the Cross Bay to Ham Lake portage. I recall he said when he got there he tried to put out the fire with a water container but said he was not able to put it out. I recall I asked him what he did next and I understood him to say he paddled away. I recall I then asked him how he got out and I recall him to say he came out over the portage. I recall I then asked him how the portage was upon his exit because I was curious if it was still burning and I recall him to say it had already burned over."

Soon after this encounter, at approximately 6:15, Deputy Tim Weitz learned that a person of interest (who was in the area where the fire began) was at Tuscarora Lodge. Deputy Weitz drove to the lodge and noticed the fire was still burning in places across the property. There were no trees torching, so it wasn't a public safety issue, but Tim recognized the area had clearly been severely impacted by fire.

When Sheriff Mark Falk learned of Steve's presence at Tuscarora, he radioed USFS law enforcement officers Barry Huber and David Spain and let them know Steve had returned. Since Barry and David had only recently discussed Steve with the USFS campers, they were definitely interested in speaking with him and hearing what he had to say.

From what Deputy Weitz knew about the fire, it had started in the Superior Forest. Technically, interviewing and questioning Steve was a turf matter. Since the fire appeared to have begun on Superior Forest land, it was under USFS jurisdiction, and not a county law enforcement issue. Deputy Weitz also recognized that the USFS had law enforcement officers trained in fire investigation. Consequently,

Deputy Weitz identified Steve, but mostly for the purposes of making sure he remained in the area until Barry and David arrived.

The first thing Deputy Weitz recognized was Steve's overall demeanor. He was "disheveled." It wasn't unusual, Tim thought, given the guy had been camping in the forest. Campers differ in how well they care for their overall appearance when in the woods. And this early in the season it was doubtful anyone would have tried to bathe in the lakes, given the ice had been out only for a week. Still, what Tim first noticed was Steve's unkempt appearance.

Because Tim knew Barry and David were on the way, he made some brief small talk with Steve, asking him where he had camped. Steve reiterated he had been down south on Cross Bay Lake. Weitz took him at his word and believed if that was the case, the guy probably didn't have anything to do with the fire and may not have been the person who was camped on Ham Lake. Deputy Weitz moved off to wait and make sure Steve didn't leave before meeting with Weitz's colleagues from the USFS.

Around 6:30 Barry and David arrived. Deputy Weitz greeted them and mentioned to Barry that he didn't think this was their guy, given his statement that he hadn't been camping on Ham Lake. But as soon as Barry saw Steve, he knew he was the person described by the USFS campers.

"I arrived at Tuscarora to find an overweight, white male, 50 to 60 years of age, with grey hair, wearing a red flannel shirt under a green 'Dartmouth' sweatshirt, looking tired, dirty and disheveled," noted Officer Huber in his statement, recalling how the meeting began. "He was identified as Mr. Stephen George Posniak. He stated he had just completed a two night camp/canoe trip on Cross Bay. Posniak provided a Wilderness Permit (#918409), which he surrendered. He stated he began Thursday May 3rd at Cross Bay Put-in, paddled through Ham over to Cross Bay Lake where he stays for two nights. He then says he was returning to the same take-out as he started when about 09:00 he sees smoke at the campsite in question, goes ashore and attempts to control the fire by dumping water with a plastic bucket, but with no effect. He stated he wanted to report the fire but had no means of communication. He leaves the fire and returns to Tuscarora Lodge where he had rented his canoe and equipment, arriving about 17:30. I asked how long it takes to paddle from Ham Lake to the take-out he said 'It takes me longer than it used to.' He said he had no contact with any others while he was out. He said he used a propane stove to cook with and never had a campfire, except when he lit a trash fire to burn off his paper trash. He assured us that he extinguished his fire with water before leaving the site, at Cross Bay Lake. He denied camping on Ham Lake, seeing or talking to anybody else during his two night trip.

"I told Posniak that I had witnesses who stated they saw him at the site where the fire started and he again denied camping on Ham Lake. I told him he was free to go but to call me that evening when he gets a hotel room in Grand Marais that night, because I need to take a photo of him. That evening I never received a call from Posniak."

After Barry ended their conversation, Steve returned to sorting through his gear in the two packs.

Barry asked Andy if he could see Steve's canoe, and Andy showed it to him.

Steve hung around Tuscarora after everyone else had evacuated. Andy was the only person remaining on the grounds. Steve was hoping to spend the night at Tuscarora, but given the evacuation order, that wouldn't be possible. When Steve wondered if he could use Tuscarora's phone to call Grand Marais and secure a room, Andy let him know their phone service was out. Finally, at around 8:15, Steve got into his SUV and drove away.

13 SPOTTING OUT OF CONTROL

After dropping off Acting District Ranger Carlene Youkum, Kurt Schierenbeck drove up the Gunflint Trail, returning to the fire's front lines. Having flown over the fire three times, he had an excellent overview of its direction and speed. What he had seen left him more unsettled than he was before his first flyover, around 2:30, when his estimate of burned forest was approximately 480 acres.

Normally, as the sun begins settling into the western horizon, the wind begins to abate. But at the time Kurt was back on the front lines, the wind showed only minimal signs of diminishing. The relative humidity was still low, and there was no sign of rain as far as the eye could see. Given the weather, they could be in for a long night.

Kurt worked with his team, particularly Vance Hazelton, Tom Lynch, and Pete Lindgren, who were keeping tabs on the fire and addressing it wherever and whenever they could. At this point, Kurt and his team had two primary goals. They needed to use whatever means possible to keep the fire from advancing to the west and north. For that effort, Kurt had a good grasp of his tools, including aircraft, fire engines, dozers, and fire personnel. They also needed to make sure that both firefighters and the public were out of the path of the fire and kept safe.

Kurt, Vance, the rest of their team, and others spent the next couple hours monitoring the fire's progress. While the aircraft were successful in bending the arc of the fire so that it pivoted around Tuscarora and most of its buildings, their efforts to snuff the fire or impede its advance were ineffective. Again, the wind, heat, drought conditions, and relative humidity were creating a perfect storm. This fire was burning hot and fast, and west of Tuscarora there was a lot of wilderness to burn.

Everyone was still hopeful that the fire's trajectory, providing the wind's direction did not shift, was taking it almost directly into the eastern center of the previous year's Cavity Lake fire, where it should be easier to manage and, ultimately, extinguish.

The other advantage to the high winds was fire width. For the moment, the winds were pushing the fire fast in one direction, which translated into a narrower burn spread. Per Kurt's earlier observation, the fire was burning almost due west in a relatively narrow cigar-like shape. Its narrow breadth should make it more easily managed as well as keep it from twisting north.

Because of darkness, the air attack had to be called off, and the air support were about ready to head back to their bases. Air attack group supervisor Jody Leidholm told the fighters on the ground that the air attack had been successful, the flanks of the fire were largely contained, and the trajectory of the fire still appeared to be largely headed west. He also thought the pace of the fire's front lines had slowed.

While Jody's prognosis was favorable, he had dispatch call up a Bell 206 B1 (3BB) helicopter out of the Hibbing tanker base and suggested it be relocated to the Grand Marais airport where—if it was needed in the morning—it wouldn't be far from the fight. The 3BB took off for Grand Marais.

The light helicopter would be used in what was referred to as a helitack. A helitack is staffed by a licensed, federally certified pilot, a helicopter manager, and two crew members. Each light helicopter is also equipped with a 110-gallon bucket. The helitack needs a landing zone, or LZ, a place typically near the fire's heel where the crew can be dropped off to fight the rear flames. The helicopter also needs a dip site, where its bucket can fetch water. During operation, the pilot lands the helicopter at a landing zone near the fire, and the two crew members get out and affix the 110-gallon bucket. The two crew members and the helicopter manager remain on the ground while the helicopter lifts off, fills its bucket, and attacks the flames.

The three ground crew have five-gallon bladders with sprayers affixed to their backs. In theory, the helicopter drops its load of water on nearby flames and the three ground crew members use their sprayers to douse and mop up any peripheral flames or embers that might have escaped the drop.

Unfortunately, the helicopters can only carry enough fuel to keep them airborne for approximately one and a half hours. The flight from Hibbing to the fire near Ham Lake would burn up half their fuel, which is why they were moving to the Grand Marais airport.

Since the Grand Marais airport did not have the proper fuel, a helicopter fuel truck was also ordered to Grand Marais. Three heavy helicopters had also been called up: Helicopter 697, a Sikorsky S-61, and a Sikorsky S-64. These heavy helicopters are used primarily for external cargo and water bucket operations. For example, the S-61 can carry one thousand gallons of water or a large number of passengers.

At dusk, Jody returned to base. He was in for a long night, because after he landed in Hibbing, he would be tossing his gear into his truck and driving to a Grand Marais hotel, where he would spend the night and be close to the Grand Marais airport. If he switched his base to Grand Marais, he would be that much closer to the fire, and he felt certain he would need to get up on top of this one not long after first light.

Less than a mile up the Gunflint Trail from the Round Lake Road turnoff, there is a parking lot for the Kekekabic Trail. That trail runs approximately forty miles from Snowbank Road near Ely to the Gunflint Trail near Tuscarora. Kurt and Vance were looking for ways to better view the oncoming path of the fire and witness its behavior on the ground so that they could determine if and where they could create a control line that would break the fire's progress. They realized they could use the Kekekabic Trail to get a closer look at the fire.

Around dusk, they drove to the Kekekabic and parked. As they hiked in, the path rose away from the parking lot, heading due south. It was already smoky. They walked in for approximately a quarter mile. Kurt was hoping they would find a place where a bulldozer could clear a line through the trees. But the closer they got to the edge of the fire, the more smoke they encountered. Visibility diminished, and they began seeing pieces of burning birch bark picked up by the wind and carried over their heads to be deposited downwind, where another spot fire was born. It didn't take them long to decide it was too dangerous and difficult to continue, and they were forced to return to the lot. They also realized that with this wind carrying embers out in front of the main fire, the spotting made it impossible to put down a control line.

By the time they returned to the truck, it was almost dark. Unfortunately, the wind wasn't settling down with the setting sun. They spent the next hour or so cruising up the Gunflint Trail, searching for places along the road that might be possible areas for the bulldozers to get in and do their work. The dozers and the fire engines were trying to get into the forest west of the Kekekabic to put out spot fires, but the wind was too strong, the spotting was too rampant, and the country where the fire was burning was too rocky or swampy to find an easy way in. And it was getting dark.

When they arrived at the Seagull Guard Station, Kurt realized he needed more resources. Even though the Type II team was on the way, before they assumed command—which wouldn't happen until morning—he ordered up more firefighting crews, engines, aircraft, and miscellaneous personnel.

Vance spent the time documenting his first day's activities.

Once he finished, he turned to help manage communications using the radio.

Kurt needed to be at the Grand Marais USFS office by 10:00 p.m. to greet and brief the first members of the Type II team, many of whom would already be on-site. Since he was forced to leave the area, he needed to find the highest rated incident commander still in the area to manage the fire in his absence, until the Type II team arrived early the next morning. All afternoon he had been in and out of radio contact with Tom Lynch, who had been putting out spot fires at Tuscarora and up and down the Gunflint Trail. Tom had been close to all the activities and was well versed in the current efforts to manage the growing blaze. More importantly, Tom was a certified Type III IC. At around 9:00 p.m. Kurt contacted Tom Lynch by radio and handed over temporary command. Tom would be the incident commander until morning.

For Tom, Vance, and several others fighting the fire, it was going to be a long night. The firefighters on the ground had been heartened by Jody Leidholm's prognosis. If his prediction was accurate and continued to hold, the burn would follow standard behavior for a fire at night. For now, they were glad to hear it.

THE PALISADES, SEAGULL LAKE, LATE AFTERNOON AND EVENING—By the time Gus, Lee, and Layne pitched their tents, it was getting to be late afternoon. Even here, though largely out of the wind, the smoke fog persisted, and ash kept falling out of the sky. Overhead was the omnipresent drone of firefighting aircraft as they continued their to and fro, flying toward the fire like bees to a hive. The southern smoke wall continued to expand, and the wind, when they climbed to the top of the palisades, was still blowing strong.

At times the smoke grew so thick they couldn't see nearby Miles Island. But at other times it thinned out, and they had better visibility.

Lee, familiar with the area's natural and prescribed burns, considered the distant smoke and apparent angle of the burn. Though it was miles away and difficult to discern clearly, the smoke appeared to indicate a burn in a mostly western direction. If that trajectory was accurate and it continued, he thought it would burn straight into the Cavity Lake fire and, choked for fuel, burn itself out.

There had also been prescribed burns on Three Mile Island and across the lake to the south. Lee figured even if the wind shifted to a more northerly direction, coming out of the east-southeast, if the fire traveled far enough west before angling north, there was a good chance it would hit the prescribed burns and, again starved for fuel, fizzle out.

The acrid fog and ash continued to burn their eyes and throats and get into their food. But Lee was surprised by his ability to tolerate it. He wasn't coughing. It was definitely unpleasant, and when they blew their noses, the mucous came out black. The ash, Lee (the university professor) reasoned, was really just calcium and phosphorus, providing them with extra minerals, if you thought about it from a purely scientific perspective. It was definitely survivable.

After dark, when Gus donned his headlamp to climb to the top of the cliff, there was so much ash falling out of the sky it was hard to see. The air was so thick with ash it reminded him of bubbles in an air column while scuba diving.

Late in the day of May 5 they considered the burn from the palisades' high point. It appeared to Lee that the fire had slowed. Considering what he could see, he thought it might have hit the prescribed burn area. If that was the case and the fire burned out, they might be able to get out in the morning. There was a chance, they thought, that they might even be able to resurrect their paddle into Alpine and beyond. They would just have to wait and see.

Unfortunately, the brief optimism didn't change the darker thoughts that surfaced in the middle of the night, particularly one filled with oncoming fire. All night Lee and Layne periodically awoke, unzipped their tent fly, and made sure their campsite wasn't covered in flames.

GUNFLINT RANGER DISTRICT OFFICE, SUPERIOR NATIONAL FOREST, GRAND MARAIS, AROUND 10:00 P.M.
—For Kurt Schierenbeck, Saturday, May 5, had been a very long day. He had now been up and working for more than fourteen hours, twelve of them on one of the earliest and wildest wildfires he had ever seen. It was too early to tell whether it would be a monumental blaze, but for this early in the season, it was already roaring like a lion, and it showed no signs of doing anything but increasing that roar in the future—at least for the next twenty-four to forty-eight hours.

The difference between Type III and Type II Incident Command System management is largely a matter of degrees. All of the resources used to fight a wildfire from a Type III perspective are also used in a Type II approach—just a lot more of them. And because of the increased size of the effort, there is a change in the way Type II incident commanders manage the firefighting.

Kurt—and now Tom Lynch—had not only managed their resources and those of others working on the fire, but they had also been on the ground. In essence, Kurt and Tom had walked the line between managers and hands-on participants. Type II efforts involve a senior management team, the duties for which are referred to as CFLOPS: Command, Finance, Logistics, Operations, Planning, and Safety. Type II commanders have people to lead each of these areas. And they in turn have personnel and equipment resources on which to draw, some more than others, depending on their roles.

During the first few hours of the fire, Kurt and his team were managing everything. Once the Type II team takes over, the Type II IC works with his core team of leaders, and they manage everything.

"The incident commander [IC] in a Type II fire isn't the one saying we need this many dozers, shot crews, and more," explained Tim Norman, USFS forest management officer, retired. "The OSC [Operations Section Chief] is dictating tactics and strategy for the incident. It's all approved or disapproved by the IC, who is acting as a CEO. The IC is managing the organization. Every one of those [CFLOPS] functional areas are feeding information into him. He or she is on the conference calls, and interfacing with the forest supervisor and the district rangers of the world."

Now there would be a larger number of resources devoted to fighting the fire. Now they would need to create a base camp, and they would require significant logistical support, not to mention having someone manage the finances, day-to-day operations, and more. But before any of that could happen, Kurt needed to inform them about all aspects of the fire and the fight, at least what had happened to date.

In Grand Marais there was a core group of thirty waiting in the Gunflint Ranger District Office conference room. By the time Kurt walked into the room, ready to brief the Type II team, John Stegmeir and several of his colleagues were present, as well as Sheriff Mark Falk, Tofte District Ranger John Wytanis, Gunflint District Ranger Carlene Youkum, county and state officials, DNR staff, and other USFS department personnel.

Kurt had been in this situation before. He knew he needed to keep the briefing to no more than thirty minutes, including ten to fifteen minutes for questions. First, he told them about the current status of the fire. He told them that at 2:30 p.m. the fire had burned approximately 480 acres, and now he estimated it was closer to 1,500. He explained that it was a running crown fire, with dramatic torching and spotting out in front of the main burn, in large part because of the weather. Weather, of course, had played a huge role in the fire's rapid progress, so he talked about it, including both the current and predicted forecast. Then he covered the personnel and resources, both USFS and all others, that had been used to fight the fire so far. His detailed list included the following:

2 USFS Type 6 engines with personnel

3 GTVFD engines with personnel

1 DNR Type 6 engine with personnel

2 Type 3 dozers with operators and leadership

7 misc. USFS personnel (including myself)

USFS law enforcement

Cook County law enforcement

He explained that they had already used substantial air resources in the fight, and he provided a list of the aircraft used the first day:

2 CL-215s

2 CL-415s from Canada

1 P3 air tanker

2 Beavers

2 Type 1 helicopters

In addition, he also told them what resources were on order.

He had a long list of items to cover. Appendix D of the Redbook listed four pages of items that should be covered, including everything from the "Incident Name," to the "Overhead and Suppression Resources Currently on Incident and Present IC," to "Fuel Types, Topography, Fire Behavior." Kurt covered each of these items in rapid detail, running down the list. Appendix D also had sections that covered "Air Operations," "Environmental, Social, Political, Economic, and Cultural Resource Considerations," and "Communications." Obviously, Kurt covered the air operations and communications in detail.

"But there are a lot of sections in Appendix D that are, at this point, just not applicable," Kurt explained. "At that point there were no procurement agreements to worry about. No injuries. No tribal lands were affected. And there were no infrared night flights. So I skipped those sections."

One point he was careful to cover was *training*. "When you delegate fire to an incoming team, training is important," Kurt said. In the Gunflint District they had several trainees who needed experience in everything from fire planning to logistics to aviation to operations. For USFS professionals to get the practical experience they needed on their way to getting their Red Cards, they needed to work with key people on the incoming Type II team. Kurt spent a few moments introducing these trainees and indicating their assignments.

At this meeting John Stegmeir also received a delegation of authority to assume command of the fire from the agency administrator. That document was the official, paper handoff of the fire from the USFS to the Type II team. Tofte District Ranger John Wytanis signed off on and oversaw the process. The document articulated what the USFS would like to see happen to the fire in the Gunflint District.

After Kurt's presentation and ten to fifteen minutes of questions, everyone began formulating their plans. John Stegmeir went off with Greg Peterson, his core operations leader, and a handful of other core staff, and they began drafting an incident action plan for the following day. Kurt Schierenbeck assisted with the creation of this plan, though they all knew it would shift once the Type II team got on the ground and took over the fire, tomorrow morning around dawn.

After they finished the draft plan, the only people who headed up the Gunflint Trail were Stegmeir, Peterson, and a couple of others, who wanted to get a feel for what was going on. All the other people at the meeting either headed home (if they lived in the area), rented hotel rooms, decided to sleep in the Gunflint District Office basement, or pitched tents outside on the grounds.

Since everyone would be up and at it early in the morning, well before daybreak, they knew they needed to get whatever rest they could.

By the time Kurt drove back through Grand Marais toward home, it was after midnight. He was profoundly tired but hopeful the fire's direction and traditional quietude that normally descended with night boded well. But since the fire had covered so much ground so quickly, he just didn't know for sure what would happen.

. . .

One or two fire engines were left patrolling the Gunflint Trail. Don Kufahl spent the night at Tuscarora Lodge, putting out spot fires and watching for embers to snuff out. At one point after dark, Tom Lynch ran home to change his clothes. Because it was so early in the season and the fire had come on so rapidly, he hadn't had time for a proper pair of socks to put on under his boots. Now he had blisters.

He drove home long enough for a fresh pair of socks and then drove back to the Seagull Guard Station, where he and Vance would both spend the night. Like Kurt, they were hopeful the flanks would hold and the front of the fire would remain blunted. But so far nothing about this fire had been easy. As the dark middle of the night settled in, they could not help but feel concerned about what the dawn might bring.

Throughout the afternoon and evening, the media had covered the fire almost from that first 11:34 a.m. call. Enquiries came in to Cook County dispatch from numerous media outlets, all wanting more information about a fire they sensed might be significant. The Cavity Lake fire was still fresh in everyone's mind, especially local and statewide news outlets, and if the Ham Lake fire turned into anything like Cavity, the media would be peering through the plumes of smoke for the story.

Throughout the day Cook County fielded enquiries from WDIO (the local ABC news affiliate), WTIP North Shore Community Radio, Kare 11 news, the *Duluth News Tribune*, and others. In keeping with protocol, the MIFC in Grand Rapids issued a press release. The release communicated the name of the fire, the time it started, and its size—600 acres—at 3:00 p.m. It also stated, "the following additional resources have been ordered: one Incident Command Team, one strike team of engines, two 20-person crews, one investigation unit, and two type six engines and one type one helicopter." It concluded by stating the Duluth office of the National Weather Service had "issued a fire weather watch for northern Lake and Cook counties from Sunday morning through Sunday evening. It will be dry, windy and warm on Sunday, factors that could create the potential for severe wildfire conditions."

Sue Prom, co-owner of Voyageur Canoe Outfitters at the end of the Gunflint Trail, was also working late on what was perhaps the best personal account of what transpired on the first day of the fire. Before dropping off to sleep, she posted a detailed, in part eyewitness report of the day's activities:

> Sometime you just know when something bad is going to happen. I woke up with that strange sense of something amiss and when Mike asked about the weather I said the forecast called for 20–30 mile per hour winds the next two days. I knew with already dry conditions a couple of days of wind would only dry things out more quickly. As I drove to town I thought to myself, "I hope the winter's snowfall put out all of the fires from last year, I sure don't want another fire." A few hours later I received a call from Marilyn saying, "Have you heard about the fire?"
>
> Women's or Sue's intuition must have been working overtime today because our good friend Sue Ahrendt also had a strange feeling this morning.

She went on to describe how Tuscarora was hit:

> The USFS, DNR, Gunflint Trail Volunteer Fire Department, Arrowhead Electric and Cook County Sheriff's office were all on scene. Hot spots and snags were everywhere and crews began work immediately. Two bulldozers dug lines around the perimeter while sawyers dropped trees and cleared areas near buildings. Sprinkler systems were installed on some buildings and around the perimeter by the GTVFD. . . . Crews will be on scene throughout the evening to help protect the many structures at Tuscarora. Winds continue to blow from the East South East with gusts up to 20 miles per hour.

She continued to talk about the fight, thanking neighbors and others, and provided some snapshots of overhead planes, a burned-over section of woods, and a large smoke plume in the distance. She also described the size of the burn at approximately 1,000 acres (the actual number will be 1,504).

Sue, along with everyone else on the Gunflint Trail, was hopeful. She concluded her post: "The fire is estimated to have stopped at the [Kekekabic Trail Crossing], and did not make its way out to the Gunflint Trail."

It was the end of the first day of the Ham Lake fire. Most hoped, after the incoming Type II team got it under control, that it would be extinguished in two, maybe three more days. What no one yet knew was that the wind was about to shift, a change in direction that would have a profound effect on the area, helping this fire grow into what would become one of the largest and most devastating forest fires in Minnesota history.

FIRE DAY TWO

14 AN ALARMING GLOW

By the time John Stegmeir and Greg Peterson arrived at the Gunflint Ranger District's Seagull Guard Station, more than fifty miles up the Trail, it was after 2:00 a.m. They checked in with those at the station, but both knew they needed to get some shut-eye, because it was going to be an early start to what was certain to be a long day.

John had a long history working in fire. He started his career with the Minnesota DNR's Division of Forestry in 1975, as an area forester at Orr, Minnesota. In that capacity he was responsible for all aspects of forest management, from planting to harvesting to pest protection, including protecting the forest from fire.

The next two years in Minnesota, 1976–77, were dramatic fire years, for which the state was not properly staffed. As a result, during the next ten to fifteen years Minnesota embarked on creating a large fire management organization, well equipped and well staffed, to address whatever wildfires were burning out of control, regardless of where in the state they occurred. During this time Minnesota federal and state firefighters realized they could better manage large fires if they combined forces. And that is when the Minnesota Incident Command System (MNICS) and the Minnesota Interagency Fire Center (MIFC) were both born. MNICS pools resources from six organizations: the Bureau of Indian Affairs (BIA), the

Minnesota DNR, Minnesota Homeland and Security Management (MN HSEM), the National Park Service (NPS), the USFS, and the U.S. Fish and Wildlife Service.

The organization also adopted national standards for the training of firefighters. In that capacity John, already with several years of experience, first became a crew leader, then a division supervisor, and then a logistics section chief, a position he held from 1984 to 1998. At that time there was some natural attrition in the organization, and in 1999 John became a Type II incident commander. For the past twelve years he had worked on numerous fires, including in recent years several off the Gunflint Trail.

Similar to John, Greg Peterson, the Type II team operations chief, had a long history working in fire. He began in 1975, while he was in college. By the time he finished his degree at the University of Minnesota, Crookston, he had been hired full-time by the USFS, which is when he went through formal fire training. From 1978 to 1994 he worked in forestry recreation and fire management for the USFS. Then in 1994 he became a fire management officer with the BIA. During that time he also obtained his Red Card for a Type II operations chief, and for the past several years he had been part of John's team.

Greg, Division Supervisor Tom Roach, and some others bedded down in the bunkhouse at Seagull station. Others were in trailers, and some pitched tents. John Stegmeir parked his Explorer so the back end was facing the distant, southeastern head of the fire. By the time he bedded down, there was a dull glow in the southeast, probably more than normal for a nighttime fire, but he was hoping it had bedded down with the rest of the world. Then he spread out his bag in the back of his Explorer and stretched out to sleep.

After the briefing in Grand Marais, John and Greg both saw favorable signs in the fire's progress. The latest weather reports showed the wind was pushing the fire in the right direction. They were concerned that the wind was clearly still blowing. But they liked the fire's trajectory. Just as importantly, it would keep it out of the BWCAW, and they wouldn't have any regulatory issues related to using heavy equipment—dozers, fire engines, and more—to fight it.

The distant glow was the kind of fire activity that caused a Type II IC to rise periodically throughout the night to check on it. When John did, the fire did not appear to be diminishing. If anything, it was getting brighter. And even more alarming, it appeared to be creeping closer. And that was when he began to seriously worry about the Ham Lake fire.

THE PALISADES, SEAGULL LAKE, MORNING—In the predawn darkness Layne awakened the others and told them they better come out and have a look. Gus scrambled into his clothes, unzipped the tent, and peered toward the east. The horizon was stained red.

"That's just dawn," Gus said.

But the hour was wrong. It was too early for first light. Clearly, the fire was still burning. Somehow, the wind must have shifted enough to push the flames around the edge of the Cavity Lake fire and the prescribed burn area, and while the fire was largely starved of fuel, it had found a path with enough downed and dry tinder to sustain it. And it had crept around the border of the prescribed burn, perhaps sensing—in a kind of anthropomorphic way, or at least following the call of its own inner physics—that there was more fuel to the north, including the kiln-dried wood of resort and cabin walls.

Feeling dejected, the three began to fear the worst; the fire hadn't stalled, and it was moving to territory that included structures that would certainly burn. They would be stranded here another day to witness the incendiary rage.

Later in the morning Layne got into his kayak and paddled out of the crescent bay's protected waters, around the point, heading out into the open water of Seagull to test the wind and waves. But not long after disappearing, he returned.

"No way you guys are taking a canoe into that," Layne said. The wind was still too strong, the waves too high, and the crossing clearly too risky to make.

They were socked in for another day, their campsite and all of eastern Seagull covered by a haze that continued to burn their throats and eyes.

SEAGULL GUARD STATION, NEAR THE END OF THE GUNFLINT TRAIL, MORNING—On May 6, dawn broke a little after five o'clock. Kurt Schierenbeck was already in his truck, driving up the Gunflint Trail to check in at the Incident Command Post (ICP) at the Seagull Guard Station. Technically, the IC had been handed over to John Stegmeir and the Type II team the previous night. But people and resources were still flowing into the region, and it would be awhile before Stegmeir's entire team was on-site. So far, Kurt was the only IC who had been over the fire, so he knew his perspective would be needed.

At the station, John Stegmeir, Greg Peterson, Tom Lynch, Tom Roach, Vance Hazelton, and several other senior, junior, and newer firefighters, including volunteers and professionals, as well as county officials and others, numbering approximately a hundred, were assembled and waiting for the first morning briefing. Some, like John, had not slept well overnight, sensing or actually seeing the approaching flames from the east-southeast.

What was clear—as everyone awoke—was that the fire had not bedded down overnight, which was extremely unusual. In fact, the wind was already up and blowing, and before the day was over, the National Weather Service would issue another red flag warning; extreme fire behavior conditions would exist well into the evening hours. And though this early in the morning it was not yet clear, the wind was almost finished making a dramatic turn. Yesterday it had blown out of the east-southeast. Today it had already shifted almost ninety degrees, blowing out of the south-southeast. That shift and the continued high winds would have dire consequences for those on the ground and overhead fighting the fire, and for the residents who were living in its path, farther up the Gunflint Trail.

But for now everyone awoke and began preparing for another day of fire.

On his way up the Trail, Kurt ran into Tom Kaffine, a USFS fire-fighter. In the early morning hours of Sunday, predawn, Tom and a couple of colleagues had been tasked with evacuating a known seasonal camper, a person local residents referred to as the Kek Man. From March through November, the Kek Man camped solo in the Superior National Forest by Mine Lake. He resided approximately 2.5 miles off the Kekekabic Trail, and because he was potentially in harm's way, Tom and his colleagues were asked to evacuate him.

As it turned out, the previous day the Kek Man had seen the smoke and air activity and as a precaution had paddled out to an island on Mine Lake, where he spent the night. The fire actually traveled over and around him during the night, and that morning—pushing through a front of flames that were sometimes so hot the three rescuers had to hold their gloved hands in front of their faces to keep them from getting burned—they were able to get the Kek Man off the island and out of the woods. But it had been a dicey evacuation that required them to return to the parking lot through a swamp rather than along the Kekekabic Trail, parts of which were burning.

When Tom spoke with Kurt, he verified that the Kek Man had been evacuated but also told him about the fire's behavior. It had grown worse overnight. Kurt would have to share Tom's observations with Stegmeir and Peterson.

Others who lived in the area and were about to play critical roles in the coming hours were also heading up the Gunflint Trail. Tim Norman left his house about the same time as Kurt. Tim was driving his white USFS Dodge truck. The truck's bed was filled with everything he needed to stay up on the Trail and work on the fire for an extended period of time. "I had clothing, a tent, and my sleeping bag in a pack. I had a box of Meals Ready to Eat (MREs), water, and my fire tools. All under my topper." All fire management officers carry this kind of gear, because they must be able to go anywhere at a moment's notice and be able to stay and fight for however long it takes. Tim also had the kind of equipment in his truck that would enable him to get himself out of trouble, if necessary. "If a tree is across the road, I can grab my chain saw, cut it and get the heck out of there. You never know what you're going to get yourself into."

Technically, Tim's job was to observe and advise, particularly since he was familiar with the area. But because this was a transition day, many of the members of the Type II team were not yet on-site. And since the fire was about to become dramatically worse, Greg Peterson asked Tim if he could help out until the rest of his team arrived.

At 5:50 a.m. Sheriff Mark Falk radioed to Cook County dispatch and let them know he was in his squad car, driving up the Trail. At 5:54 Deputy Chief Leif Lunde radioed in; he was also headed up the Trail. Later that morning Deputies Doug Rude and Tim Weitz would also drive up the Trail.

Prior to the 7:00 a.m. briefing by the Type II team, Kurt and Tim met with Stegmeir, Peterson, Lynch, Roach, and others to review the previous day's activities and next steps. Judging from the wind, the approach of the fire, and its apparent growth overnight, Greg Peterson knew that right after the meeting, he wanted to view the fire from the air, so he had arranged for a Beaver out of Ely to pick them up at Blankenburg Landing on Seagull Lake.

Greg asked Kurt, Tim, and Tom Roach to accompany him on the flight. Kurt could answer questions about yesterday's effort, and could review where the fire had been and compare it to where it was now. In addition to being familiar with the area, Tim had a lot of firefighting experience and expertise; Greg wanted his advice and counsel. Tom, a seasoned firefighter who would be leading Division A on the fire, would be responsible for the section of the fire from Round Lake Road (the turnoff to Tuscarora), all the way up the Gunflint Trail to just past Seagull station.

Later that morning there would also be a Division B assigned to the fire. DNR firefighter Dan Grindy had been called up as a division supervisor but was not yet assigned. By midmorning he would be leading Division B and would be in charge of the area from just past Seagull station to where the Gunflint Trail ran by Seagull Canoe Outfitters, approximately four miles farther up the road.

Finally, when Sheriff Falk reached the station and found out there would be an overhead flight with firefighting experts, he decided to go along. He not only needed to hear the opinions of the experts in the air, but he needed to see the fire from above and figure out whether more people would need to be evacuated.

The seven o'clock briefing was attended by everyone at the ICP, including Cook County Emergency Management Director Nancy Koss, and other county officials, as well as a number of volunteer and professional firefighters. Morning briefings convey important information, including the weather forecast, planned burnouts, resource call-ups, evacuation plans, and so on. These are typically run by the IC's plans chief. The briefings seldom last more than thirty minutes and follow a set process, or checklist, for who speaks and what they need to cover. The operations chief would talk about the day's operations. The fire behavior analyst might talk about the fire. Then at the end of the meeting, the sheriff and a county representative would be given an opportunity to speak, if they had information that had to be shared. Finally, the IC would conclude the meeting.

But this was a transition day, and only some of the Type II team had arrived, so Kurt did most of the presenting. The USFS, DNR, volunteer, and other firefighters who had been and would be on the front lines of the fire had not been at Kurt's presentation to the core Type II team the previous night in Grand Marais, so he reiterated much of what he had shared with that team. He reviewed the resources they had used the day before, and those they were already using today, as well as the additional resources that had been called up. Then Greg Peterson reviewed the draft incident action plan (IAP) that he, Kurt, and John Stegmeir had worked on the previous night.

The IAP was handed out, because this was the document that told the division supervisors and everyone else what they would be doing for the day. It covered the operational period from 0700 to 1900.

The IAP's overall objectives were relatively straightforward. Because Kurt had worked on this fire and noticed, traveling early in the morning to the ICP, that the fire was growing more dangerous, the objectives were especially concerned with safety:

- PROVIDE FOR FIREFIGHTER AND PUBLIC SAFETY IS THE PRIMARY OBJECTIVE.

- PROTECT STRUCTURES AROUND TUSCARORA LODGE.

- MINIMIZE DAMAGES AND MAXIMIZE OVERALL BENEFITS OF WILDLAND FIRE WITHIN THE FRAMEWORK OF LAND USE OBJECTIVES AND RESOURCE MANAGEMENT PLANS.

- USE SAFE AND AGGRESSIVE MEANS TO CONTAIN THEN CONTROL THE FIRE.

- USE M.I.S.T. WITHIN THE BWCAW BOUNDARIES PROVIDING FOR SAFETY FIRST.

MIST stands for Minimum Impact Suppression Tactics, a term and practice with which most of those assembled were familiar.

Because professionals were still flowing into the area, the number of front-line personnel was still thin. Most of the Type II team leaders had arrived, including the IC (Stegmeir), the operations chief (Peterson), the planning chief (Jim Rupert), the finance chief (John Kelly), and the air operation branch director (Keyth Wallin).

The IAP also identified the division supervisors and a group supervisor, key leaders who would have teams and engines to fight the front lines of the fire. Normally for a fire this size there would be several, but since this was still a transition day and early in the season, the IAP listed two: Tom Lynch and Tom Roach. In addition to being a Type III IC, Lynch had a Red Card for structure protection. Consequently, he was now assigned to Greg Peterson as a group supervisor on structure protection. Dan Grindy would be the third division supervisor, assigned later in the morning.

The Red Card in structure protection meant Lynch had completed enough class time and had enough experience saving structures to have met the national standards required to obtain his card. In fact, Lynch had written the structure protection plans in the Alpine Lake fire, the Cavity Lake fire, and others.

Most division supervisors were assigned to geographic areas. The regions for Roach and Grindy were, or soon would be, clearly delineated. They would be leading the firefight in those locations. Since Lynch was a division supervisor on structure protection, he would go wherever the operations chief and others believed he would be most effective.

Each division supervisor had orders and resources to draw upon for carrying out those orders. Roach's orders were to "construct dozer lines as directed by division supervisors," and "secure fireline from campsite to Cross River Landing." His initial resources, at least until more people and engines arrived, were a crew of four firefighters with a Type 6 engine managed by Vance Hazelton, and five dozers (bulldozers) he could use to create control lines.

Lynch was starting his day with four Type 6 engines. Two of the engines were from the GTVFD. Each engine had two people assigned to it. Engine 573 was led by Don Kufahl, and Engine 563 by Bob Baker. USFS firefighter Pete Lindgren led a three-person staff on Engine 22, and USFS firefighter Peter Igoe led a two-person team on Engine 21. Lynch's orders were designated as "structure protection as directed by group supervisor."

The IAP also identified people and tasks devoted to "public safety" efforts. In this case it listed Tom Kaffine's evacuation effort of the Kek Man, and Jim Wiinanen and Gene Dressley, who would be arriving at the ICP at 8:30 a.m. to paddle into Brandt Lake, where some campers needed to be evacuated. The public safety effort also included closing the Kekekabic Trail from the east entry point, off the Gunflint Trail, all the way to Missing Link Lake.

Finally, there was no, or limited, phone service in the area. The landline phones were down. Thankfully, nearly everyone had radios.

But there were numerous frequencies, and not everyone had a radio that could communicate with the assigned frequency. Importantly, the IAP included an Incident Radio Communications Plan that listed all the frequencies that would be used for air to air, air to ground, incident command, the two divisions, and so on. Judging from the flames and smoke already permeating the nearby forest, communication between all parties would be critical.

The IAP was not a static document. It was a blueprint for a logical course of action based on what was known about the fire the night before. And while the divisions would not change, the work they were assigned would, as the day progressed, reflect the chaos that was about to unfold.

Not long after the meeting concluded, Peterson, Schierenbeck, Weitz, Roach, and Sheriff Falk got into their vehicles and drove to Blankenburg Landing. Kurt was going to see what had happened overnight, and his colleagues were going to get their first overhead glimpse of the flames.

USFS firefighter Pete Lindgren was familiar with the Seagull Guard Station. In the past, Pete had participated, along with Tom Lynch and others, on conducting burnouts of the area around the station. The purpose of these burnouts was to reduce the available fuel around the station's perimeter in case fire threatened. At least once a year, weather and the conditions permitting, they burned out 360 degrees around the station's buildings. They also did burnouts across the road at the heliport and around a nearby area they called the Boneyard, an open area where there were picnic tables, boats, and canoes used for firefighting, and other types of equipment.

The station had also been set up with a sprinkler system. The system was not operational this early in the year. The hoses and sprinklers and related equipment had all been set up, but now the pump, which was stationed by nearby Larch Creek, needed to be set up with a fuel source and started. Pete got the fuel and worked his way down to the pump, situated near a large pool close to the Gunflint. He worked for a while to get the pump set up and operational and then turned the system on.

The burnout efforts to reduce fuel and Pete's efforts to get the sprinkler system operational would be fortuitous as the day and fire progressed.

"Without that sprinkler system," commented Tom Roach, remembering how the afternoon and then night had unfolded, "the Guard Station would have burned."

By now people who lived in the oncoming path of the fire were becoming concerned. While the flames and smoke were still in the distance, their path was painfully obvious to everyone in front of it.

At 7:34 a.m. Cook County dispatch received a call from someone on Seagull Lake Road, who reported, "they are getting a lot of heavy smoke, ash and black not completely burned embers—CC 10–5 info to 102 at ICP." (Dispatch forwarded the information to Deputy Chief Leif Lunde at the ICP at Seagull Guard Station.)

Pilot Wayne Erickson had flown in plenty of bad weather, and while this morning was not the worst, it was definitely not calm. The wind was blowing steadily and gusting up to thirty miles per hour, causing whitecaps to break across Seagull Lake. And the fire itself was creating turbulent air.

It took awhile for everyone to get to the landing. At 9:04 a.m. Sheriff Falk radioed dispatch and let them know he was heading into the air to take a look at the fire. The dispatcher in Grand Marais logged the call: "10–6 for recon flight off the air for a bit." ("10–6" is ten-code for "will be out of the area.")

Because Kurt had already been over the fire, he took the jump seat in the back of the plane. Beavers normally cruise at approximately 140 miles per hour, and they can carry only six passengers. Those passengers need to climb up and into seats through a small side door. Kurt had to climb over the front and back seats and stow himself in the back, where cargo typically rides. He could peer through a small portal in the back of the plane's fuselage. The tail section was also the bumpiest ride on the entire craft. Kurt was about to be buffeted around like a kid on a state fair amusement ride.

The others had more stable seats and better viewpoints out the side of the plane, but what they noticed most was the wind. Once they rose off Seagull's waters, the plane was rough in the air. Kurt especially was feeling the rocky ride. Peering out of one of his nearby portals, he also noticed that when the plane was turned south, its progress seemed to stop. When he looked down at some large boulders sticking out of the lake, they didn't appear to be receding behind them as they normally would. They appeared to be a fixed part of the landscape.

Kurt had been on all kinds of aircraft across America, including helicopters, tankers, and a variety of other fixed-wing aircraft. But he had never been on a ride quite as rough as this one, sitting atop the jump seat of a Beaver. When they finally got near the fire's front, Kurt and his colleagues were shocked by what they saw.

Earlier, in Grand Marais, Jody Leidholm was getting ready to take off. Like the Type II team, the air attack people usually had a morning briefing. Because Jody was the most senior person who had been on the fire, he shared the details about yesterday's air support and attack, including what they had called up and what was being used to fight the fire.

By now Keyth Wallin, the Type II air operation branch director, was at the Grand Marais airport, along with some news reporters. Jody knew some of the media pilots because he taught classes in the Twin Cities about how air attacks on fires worked, so they could stay out of harm's way when covering a fire from the air.

In fact, he was being interviewed by a Channel 10 reporter and a cameraman from Duluth when he checked his watch and realized he needed to get in the air.

"Any chance we can come along?" the reporter asked.

"I don't have any issues with that," Jody said. "But you need to

get approval from the Fire Center because I don't have the authority to allow it." Besides, Jody knew how rough it could get in the air,
and he could already see and feel the day's wind. "And keep in mind
you might not see the ground for five or six hours," he added, letting
them know the Queen Air did not have a bathroom.

The reporter and cameraman decided to stay on the ground.

Since the Beechcraft Queen Air cruised at nearly two hundred
miles per hour, it didn't take long for Jody and his pilot to close the
distance between the airport and the fire. He, too, was shocked by
what he saw.

"When I get out there, I don't even recognize it because the fire
had grown so much overnight," Jody said.

When he arrived at the area where he had left off the night before, he expected to find the front of the fire nearby. The fire had
not only grown much larger than he expected, but the head of the
fire had shifted. He could see that after dark the head of the fire had
pushed much farther west than he thought it would, traveling more
than two miles. Then after midnight the unthinkable had happened.
Gradually, the wind had shifted from pushing the fire west to pushing it north-northwest. The winds had rotated almost 90 degrees.
Now the entire two-mile-plus right flank of the fire was turning into
the fire's head. And if that trajectory continued, it was only a matter
of time before it neared or crossed the Gunflint Trail and began to
threaten structures and people.

Jody radioed Greg Peterson to discuss the change in the fire's
size and behavior.

"This thing has grown many hundreds if not thousands of acres.
I'm going to need a little time to map things out."

Peterson was in the Beaver with his team, doing the same thing.
They agreed to reconnect after their reconnaissance flights.

Jody pulled out his Superior National Forest map, examined the
fire area from the air, and was "blown away" by what he saw.

"I didn't recognize anything that we had done the day before because it had gotten so much bigger. I estimated the fire on Saturday night at maybe 750 to 800 acres. After I mapped it in the morning, it had grown to about 7,000 acres."

"That fire behavior," continued Jody, "is unheard of for that time of year. Fires usually don't burn well in early May at night, especially way up on the Arrowhead region of Minnesota."

But clearly, this one had.

Then he ran through his checklist of items he knew he would need to share with the numerous parties involved with fighting the fire. In fact, Jody had done this exercise often enough to have created his own checklist of subjects to note and document. He knew from experience the kinds of information each group needed. Over the next several minutes he provided the MIFC, dispatch, and John Stegmeir, the IC, with the relevant parts of each of the following fire categories:

APPROXIMATE # ACRES: 7,000 and growing fast.

FIRE BEHAVIOR, for both the flanks and head of the fire: A "crowning/spotting" fire, the most intense.

FUEL: The fire had numerous fuel types to feed upon, including grass, regen (hardwood and conifer), aspen, hardwoods, conifers (pine, spruce, lowland spruce), and slash (hardwood and conifer).

SPREAD POTENTIAL AND DIRECTION: It was spreading fast, showing no signs of abatement, and was turning north.

VALUES THREATENED: There were people, structures and lots of forest in its path.

Other fire considerations Jody covered included "Containment Chance," "Estimated Time Commitment for Tankers/Atgs," "% Knockdown/Contained," "Response to Control Efforts," and "Structures Threatened." None of what he related to any of his constituents was positive. More specifically, for Greg Peterson and John Stegmeir at ICP, he covered "Strategy/Tactics Recommendations," and "Additional Resources Needed or Recommended (air or ground)."

For each of the preceding relevant categories, the news was bad. There was much more fuel than he previously had thought. Not only had the fire shifted toward areas that contained more fuel, but the winter drought conditions meant snow had not bedded down the grasses and marshland as it would have during normal years. The head of the fire was now nearly two miles long, making the spread potential extremely high, especially given the turbulent winds. And for both "Values Threatened" and "Structures Threatened" the news was particularly bad. Unless the apparent path of the fire changed, the number of structures that could be burned could be catastrophic. And of course anyone who resided in the area, particularly in the path of this crowning and spotting fire, was in danger. They needed to evacuate.

The fact that it was only May 6 was both a blessing and a curse.

"It's so early in the season the only people at the end of the Trail are permanent residents or resort owners," observed Tim Norman. The blessing: there were far fewer people threatened or who needed to be evacuated than there would have been in just one week, with the opening day of fishing season just six days off. The curse: it was so early many of the cabin owners had not yet opened their cabins. And those with sprinkler systems had not yet had a chance to set them up and make them operational.

From a firefighting staffing perspective, it was also a curse because fires of this intensity don't usually happen this early in the year. Consequently, the USFS, the DNR, the BIA, the GTVFD, and others were often more concerned with early-season training activity than fighting a live fire.

"None of us had our full complement of seasonal workforce on hand yet," noted Tim Norman. And with the fire heading north, there was a very real likelihood it would cross over the border into Canada. The farther north you go, the later the season is, so Tim

believed that if and when the fire jumped into Canada, the Cana-
dians would be less farther along staffing their seasonal workforce
than the Americans.

One of Jody's first actions was to order up three CL-215s and
a P-3 Orion heavy air tanker, to get them going on the fire immediately. Then he began looking for ways to knock down, stop, or control
the fire. While he had the pilot fly him around the fire perimeter—
with the Superior National Forest map on his lap so he could size
it up—Jody also tried to "get an idea about what chances we might
have to do something." He was looking for roads, highways, waterways, or anything he could use, or "we might be able to tie into and
use for a line."

Jody worked with both Greg Peterson and John Stegmeir to
come up with a general plan for the day. The mission, as Jody saw it,
was to hold the fire south of the Gunflint Trail. "To hopefully keep
it south and west of the highway." He would also be doing individual
structure protection, but he doubted—given the fire and wind—he
would have any ability to do anything with the head of the fire.

"All we could do was try and check it the best we could and
save whatever we could," recalled Jody. "There were so many people down on the ground who had no idea where the fire was. They
thought they were maybe at the end of the fire, and heck no it was
a mile west of them already, and it was spotting an eighth of a mile
to a quarter mile ahead of itself. I remember talking to my pilot and
saying, 'If somebody gets killed today, it's not going to surprise me.
It's chaos down there.'"

Also overhead were his colleagues in the Beaver plane. They
were trying to get a sense of the fire's perimeter and figure out an
action plan for the day, but they, too, were seeing the same blazing
and smoky landscape as Jody.

"We're seeing extreme fire and spotting," noted Tim Norman.
He and Greg Peterson watched the fires below, spotting well out
ahead of the main fire line.

When Greg saw the location of the spotting, between the Blankenburg sand pit and Seagull's eastern shore, he thought they might be able to stop it. If they could stop it there, they might be able to prevent it from moving farther north, toward an area dotted with lakeside cabins and outfitters.

"Do you know how we can get into that place?" Greg asked.

Tim noted the location and knew once they were on the ground, he could show Greg how they could get in—possibly with engines— and start fighting the flames.

Sheriff Falk listened to the firefighting experts, and what he heard and saw alarmed him. He could see the fire was heading straight toward structures and people, and it appeared to be only a matter of time before it would cross the Gunflint Trail. He could also see firebrands being picked up by the wind and carried well out in front of the primary fire perimeter, starting fires of their own. That kind of wind and fire behavior made it doubtful the two-lane black-top of the Gunflint Trail, which traveled mostly east to west with oc-casional turns north, would create a wide enough break to stop the flames. It appeared to Sheriff Falk as though he was going to have to declare an evacuation of the area west and north of the Seagull Guard Station. From the air he could see that involved many cabins, some resorts, and lots of other structures and equipment.

Back in the jump seat Kurt Schierenbeck was still being rocked by the extreme weather. In fact, everyone on the Beaver—even those in the more stable front of the plane—were feeling queasy from the bumpy ride. And these were professionals, like Kurt, who had spent a lot of time in small aircraft in turbulent skies.

For the first and only time in his life, Kurt grew sick in the air-craft. It was just another painful indicator of how crazy the wind was blowing. He was thankful the Beaver was well provisioned with motion sickness bags.

Once on the ground, Sheriff Falk radioed dispatch and let them know he was back in service and able to take calls. At 10:09 a.m., dispatch noted the call and entered the ten-code "10–8"—meaning the sheriff could be assigned to calls. Then Sheriff Falk drove to the ICP at the Seagull Guard Station.

Everyone except Tim Norman and Greg Peterson headed back to the station. When they were up in the Beaver watching the extreme fire behavior and spotting, Tim had made a note of how to get into the area where he and Greg thought they might be able to do a burnout. They headed up the Trail to Blankenburg Lane and turned in to have a look. There was even some spotting there, this far north. But it appeared safe enough that they might be able to do a burn. Tim and Greg discussed the possibility and then decided to give it a try. Tim remained in the area, radioing Tom Lynch for assistance with the burnout, while Greg drove to Blankenburg Landing.

The previous day both Tim and Greg had feared the fire might require everything in their arsenal to fight it. They both decided it would be a good idea to put USFS Aerial Ignition Firing Boss Barb Thompson on call, just in case. Now that they were seeing radical spotting and how crazy this fire was becoming, they were glad they had called her up. Greg Peterson was heading to the landing to pick her up. It would be awhile before her skill set was needed. But when she was finally deployed, she would impact the fire in ways no one could yet foresee.

15

THE EVACUATION MOVES FORWARD
AND A BURNOUT GOES AWRY

At 10:34 a.m. Sheriff Falk finally made the formal evacuation call.

"Doing a mandatory evac from Seagull Trail north. Spoke with 102 [Chief Deputy Leif Lunde] re: printing the evac lists," Cook County dispatch reported.

Also at 10:34 Cook County dispatch received a concerned call from the public. "Reported wood pile on fire where Seagull Creek crosses the Gunflint Trail." Dispatch relayed the information to Sheriff Falk, who assured them he was aware of it.

Seagull Creek crosses the Gunflint Trail approximately one mile due west of the station, so anyone traveling down the Gunflint toward the station and beyond saw the flames threatening the road. The fire had not yet crossed the Gunflint Trail, but everyone on or near the fire sensed it was only a matter of time. The fire was out of control. The wind was strong, and the fire front was amorphous and moving, as the wind carried embers far out ahead of the main fire activity, sparking flames wherever they landed. It was difficult to know where to try to contain it. There were lots of firefighters working on the fire, but because of the extreme conditions, everything felt chaotic.

At 10:39 a.m. Sheriff Falk made another call to dispatch. "Request CC page GTVFD to assist with the evac—have them meet at the ICP at the Seagull Guard Station—CC paged 3 times."

Shortly after this call Mike Prom, GTVFD assistant chief, arrived at the ICP and located Sheriff Falk.

Sheriff Falk hated calling an evacuation order, but knew his primary responsibility was keeping the citizens of Cook County safe. From what he had seen, the only way he could ensure their safety was to get them out of the path of the oncoming flames.

"Can you evacuate everyone from Round Lake Road to the end of the Gunflint?" the sheriff asked. He knew there was a process in place. He knew there was a plan. And he knew the key players who were supposed to work that plan, particularly when it came to the evacuation.

Fortunately, the GTVFD had been practicing evacuations of the Gunflint Trail at least annually since the 1999 blowdown. Coincidentally, their practice sessions usually took place in early May, so they were familiar with implementing the process at this time of year.

"We can do it," Mike said, because Mike knew that once the sheriff called for an evacuation, the GTVFD was responsible for executing it. Cook County law enforcement already had the sheriff and several deputies in the area, and they could assist when needed. But their primary duty was to erect and man the traffic barricades and help manage traffic on the Trail.

As soon as Mike received the order, he paged all available volunteer firefighters. "The sheriff has issued an evacuation order for the Seagull Sag Zone. All available firefighters report to the Seagull Guard Station immediately."

Tom Lynch, who had several volunteer firefighters on his team, was Mike's co-leader on the evacuation. Whenever possible, some or all of his team members would assist with the evacuation.

Right after Tom's team got their evacuation lists from Mike, Tom received the radio call from Tim Norman regarding the possible burnout they wanted to do at Blankenburg Pit, back up the road.

Tim would need Tom's assistance with the effort and told Tom to meet him on Blankenburg Lane.

It was odd to get direction from Tim Norman. The call should have come from Greg Peterson. But Tom knew it was a transition day, and this fire was showing signs of burning out of control. This morning the winds were crazy. Many of Stegmeir's Type II team had not yet arrived. He had already ordered up two twenty-person crews, and they were on their way but still not on-site. Besides, Tom knew that Tim, whom Greg had asked to assist, was particularly familiar with the area. And Tom had worked with Tim in the past and knew he had great fire instincts and knowledge.

Tom was familiar with Blankenburg Lane and knew where to meet Tim. As group supervisor, Tom had traded in his Type 6 Engine for a Chevy half-ton DNR pickup. As Tim did, Tom carried plenty of equipment in the back. He and his engines were on their way, but it would take them more time than normal to get there. The flames were approaching, and thick smoke was out ahead of the front, reducing visibility.

Back on Blankenburg Lane, Tim waited for Tom. It hadn't been more than fifteen minutes since he made the call, but what he increasingly noticed was red flag weather; the day was growing windier and hotter, and what he was seeing—here, well ahead of the main fire line—gave him pause.

At 10:46 a.m. Sheriff Falk radioed Cook County dispatch. "10–8 Prom will be in charge of the evac."

It did not take long to assemble the rest of the personnel needed to implement the sheriff's order. In fact, Mike Prom, Bob Baker, and others had not only practiced the plan but had worked with Cook County and others to create it. After the 1999 blowdown one of their first efforts was to work with the county to get a complete list of all property owners on the Gunflint Trail. After analyzing that list, they could see there were concentrations of people at the end of the Trail, near Gunflint Lake, down the road at Poplar Lake, and so on. They knew that if and when they needed to evacuate people, it would probably be in different sections of the Trail; it was doubtful the entire 55-plus-mile highway would need to be evacuated at once. They also knew they wanted to be able to get people out in no more than two hours.

Mike, Bob, Dan Baumann, and others had practiced this evacuation during off-season and during busier times of the year. During off-season (Labor Day to fishing opener), fewer people were residing along the Gunflint Trail. From fishing opener through the summer, the influx in population complicated and prolonged the evacuation.

"We need to stop and speak with people," explained Mike. "The more people, the longer you talk, and the longer the evacuation takes." But because Mike had practiced this evacuation during the off-season, he was confident they would be able to get everyone out within the two-hour goal.

Given the preceding requirements, Cook County officials, the GTVFD, and others divided the entire Trail into several zones. One of those zones was called the Seagull Sag zone, the area from around Tuscarora to the end of the Gunflint Trail.

The authorities had anticipated that in the event an evacuation was called, they would need to have immediate access to current lists of the residents in each zone. The need for immediate access prompted the three GTVFD stations (Poplar, Gunflint, and Seagull) to store boxes with binders for each zone. Each zone binder at each station contained two copies of the list of residents for that zone. Earlier, Mike had placed the binder with the lists of zone residents into his personal vehicle, which he had been driving all morning. He brought that binder with him when he went to the Seagull Guard Station.

At the ICP, there was a loudspeaker on the outside of the building used to broadcast important information about the fire, the weather, and similar news. The parking area was also busy with vehicles and people. There were overhead planes droning and making drops. The first thing Mike did was move to the corner of the station, where he set up an impromptu evacuation command post on top of a garbage can. Here he spread his map and binder and began to parse out the separate sheets he would distribute to the teams heading out to notify residents.

At 11:00 a.m. Dan Baumann also radioed Cook County dispatch. "Page the fire chiefs from Maple Hill, Grand Marais and Colvill. Let them know there is an evacuation in process and please be on standby in case assistance is needed."

Dispatch did as Dan requested and noted it in the log: "CC page all three chiefs 2x."

Even if they didn't use the additional fire engines in the evacuation, Dan and Mike knew that they might very well need them to fight the fire as the day progressed.

When there were enough people assembled to begin the evacuation, Mike began distributing the lists.

"It was fluid," Mike recalled. "I distributed the lists of residences the furthest out in the zone first. As more volunteers showed up, I gave them lists of residences who lived nearer." During the whole time he was consulting his map to determine which lists should be distributed when and to whom.

While Mike was the GTVFD firefighter leading the evacuation, others were helping out. There was someone working the logistics, another colleague responsible for radio communications, and so on. Dan was shadowing Mike and assisting whenever needed.

Everyone participating in the evacuation was assigned a list, and the lists were divided in a way that made them all roughly equal in the amount of time needed to notify residents in the affected area. So evacuation workers who needed to notify people at the end of the Trail had shorter lists than others who were notifying people nearby. Similarly, roads that left the Trail and then branched out into two to three other side roads were given teams—depending on the number of residents down each road.

While Tom Lynch and his team of five engines had left to meet Tim Norman down Blankenburg Lane, when they were finished with their burnout, they would also be helping out with the evacuation. Tom had years of experience fighting fires, and it was largely Tom who trained in Mike, Bob, and others on how the evacuation should be executed. The first step in structure protection, Tom knew—particularly in a region as densely wooded as this part of Minnesota—was making sure you knew the location of every structure on the Trail.

Back at his garbage can Mike explained the process to his initial group of volunteers.

"Find the residence. If homeowners are there, let them know they need to get out and go down the Trail. If they aren't home, post an evacuation order where they'll see it, on their front door or side door or wherever."

Mike explained that on roads and turnouts where it was necessary, they needed to post signs pointing the way out. There were lots of side roads and driveways off the Gunflint. The authorities had anticipated that in the case of a fire, smoke or nightfall or both might impede the ability for evacuees to find the right way out.

Mike reminded them that law enforcement and the GTVFD cannot force people to evacuate. But they can impress on them the gravity of the situation and trust most, if not all, of Cook County's citizens were smart enough to heed their advice.

"In the instances when people are reluctant to leave," noted GTVFD firefighter Bob Baker, "we ask them to give us the name of a next of kin to notify in case something bad happens." Bob noted that simply asking the question was often enough to convey to evacuees the gravity of their situations. Usually, once they answered the question about next of kin, they were beginning to change their minds about remaining in their cabin or home.

Once a cabin or residence had been checked, each person with a list was ordered to radio ICP, where the lead communications person worked with others to track everyone's progress, noting every instance when a resident had been notified or—if no one was home—an evacuation order posted.

Over the next hour and a half the volunteers fanned out and worked down their lists, and the few people who were in the Seagull Sag Zone began to gather their personal belongings and get out. While everyone was notified well within the anticipated two-hour notification goal, some residents took longer than others. As the fire continued to blow out of control, the tardiness with which some evacuees responded to the fire put them in danger, for a time, of having their exits blocked. The Gunflint Trail ended at water. If the fire covered over the Trail farther south and east, making it too dangerous to traverse, the authorities were going to have to figure out how to get everyone out over the water.

But for the moment, Mike knew nothing about how the fire, as the day progressed, was going to make a run that would begin devouring everything in its path.

By the time Tom Lynch arrived at the Blankenburg Pit turnoff, his Type 6 fire engines were close behind. His process was to go in and examine the area and make sure it was safe for his engines to work. He looked not only at the fire behavior but also for places where he thought they could put down a line or try a burnout. He needed to know that if the fire started getting the upper hand and the area became dangerous, he and his engines had a way to escape the flames.

When you are heading toward the end of the Gunflint Trail, the Blankenburg Pit turnoff is a sharp left after the Trail begins turning north. Tom turned into the road and headed in a southeasterly direction. As soon as he turned off the Trail, he began seeing spotting ahead of the flames. Around a hundred yards down the road, there is a turn. Tim Norman was waiting for him at the turn.

There was a snowmobile trail nearby. As Tom's engines entered the area, he and Tim talked about creating a fire line along the snowmobile trail that could act as a fire-stop. If they could drive that fire west, they could burn out the area in front of the approaching flames.

As Tom and Tim were planning, fires continued spotting all around them. Tom needed time to get everyone in place to do the burnout, so he contacted Jody Leidholm overhead. He requested some air support to tamp down the nearby flames and put a damper on the spotting. Tom's full team was supposed to be in on his strategy and planning discussion with Tim, but there was so much spotting, two of his team's engines were fully occupied dousing the flare-ups.

"We were getting spotting way beyond what the analysts predicted," Tom said. "With regard to spotting there is something called the probability of ignition, or POI. In order for fire spotting to occur, you need an ember, a transport medium, and a receptive field. The birch bark firebrands were perfect embers, and the wind was a perfect transport medium, and the area was so dry anywhere the ember landed it ignited. We could do the math on the percentage POI, but we didn't need to, because it was 100 percent. Everything that landed caught fire."

Tom was instructing Jody where to drop water. But the area was so thick with smoke, Jody couldn't find them. Tom told him to follow the Gunflint Trail from the Seagull Guard Station and then beyond to the first left in the road, about a mile and a half. But neither Jody in his Queen Air or his sky crane tanker could locate them. There was still too much smoke.

Tom, frustrated, got into his truck and figured out the GPS coordinates, which he then related to Jody in the air. But it took too much time, and while he was figuring out his GPS coordinates, the fire wasn't waiting.

In the middle of the briefing, Greg Peterson arrived on the lane with Barb Thompson. Now the four of them discussed putting in some dozer lines.

Tom called the ICP to see if there were dozers gassed and ready. There were, but it was an informational call, because Tom was not yet sure they were going to have time to use the dozers. He was just trying to determine whether they were available if and when he needed them.

At the ICP there was confusion about Tom's request. Two dozers were sent down the Trail toward Lynch, thinking they had been called up.

"I remember pulling into that gravel pit with Tom Lynch and his engines there," recalled Barb Thompson. "I remember being surrounded by fire and thinking it was crazy fire behavior."

As the spotting and fire continued, the air grew heavier with smoke and ash. The overhead air attack still could not find them. They began to realize the landscape and fire were becoming too threatening to continue the operation. The area was becoming too compromised. Because the wind was so strong, spotting was happening more than a quarter mile ahead of the flames. It was a dangerous situation, and they needed to get out. Immediately.

While they were deciding to leave, a dozer on a truck and lowboy turned off the Trail onto the lane and drove a hundred yards up the road, turning at the corner. Tom halted it at the turn, realizing the lane was now blocked, with his team and their vehicles behind him, ready to get out. The fire was spotting all around them.

Tom was suddenly worried and angry. There were ditches on either side of the narrow road. If the dozer transport rig went into a ditch and jackknifed, five engines and three trucks could be in jeopardy, not to mention their drivers and operators. He had not ordered the dozer, and now the huge machine was blocking their exit. Some of the newer firefighters were clearly alarmed.

It was not easy to back up a loaded truck and trailer, but somehow the operator managed to do it without getting stuck.

Then Tom waved his engines through and told them to go to the end of the Trail, where they could regroup and start working on the evacuation, and to douse any spot fires along the way. Tom knew embers were already landing and firing far up ahead.

The engines all managed to exit. Tim Norman and Barb Thompson left. And then it was just Tom and Greg Peterson.

Greg was in his truck, ready to cover the one hundred yards of Blankenburg Pit road to the Gunflint Trail turnoff. Now it was so smoky, he could not see the Trail at the end of the lane.

Greg stopped, looked at Tom, and said, "We're going to lose some homes today." It was a fateful, dire comment. But given the chaos they were now experiencing, it was difficult to imagine the day progressing without structures being lost. Greg had been fighting fires for many years, but he was not sure he had ever seen a fire so out of control near so many structures.

Tom knew it was true. "Yes," he finally said.

"But we are not going to lose people," Greg said. When it comes to fighting fires, Greg and Tom were both aware of the USFS mantra: Life, property, resources . . . in that order.

Tom didn't have to think about it. He knew he would never place his people in harm's way, and he was always careful about the jobs they were tasked to do. He would ask of them 110 percent, but he would never place them in situations without a clear means of escape, should the tables turn. "Not on my watch," he said.

By the time Greg and Tom were on the Gunflint, Greg turning right to return to the ICP, Tom turning left to head up the Trail with his engines, it was near 11:00 a.m. At that moment it seemed as though the fire shifted up a notch, and the smoke plume that had been rising over the flames appeared to double in size.

For the next hour, everyone helping on the evacuation was driving up lanes, turning into driveways, getting out of their vehicles, knocking on doors, telling residents they needed to get out. Once they cleared a residence, they radioed it to Mike Prom at the ICP.

At 11:37 a.m. someone called Cook County dispatch. "Reporting End of the Road and the campgrounds have been cleared—CC 10–5 to GUNFD IC and 10–5 back to BP25 to return to ICP."

At 11:45 a.m. Dan Baumann radioed dispatch, changing his

previous "standby" order to an urgent request for assistance. "IC [Incident Command] request CC page MHFD, COLF, GMFD [Maple Hill, Colvill and Grand Marais] for mutual aid for structure protection—need people and equipment—also when can request mutual aid from all other departments anything they can do—CC pages 2x COLFD, MHFD, GMFD."

In addition to being the current GTVFD chief, Dan was co-founder and president of the Fire Chief's Mutual Aid Fire Association. As president, he requested mutual aid from all the volunteer fire departments in Cook County. From his work on the GTVFD and with the association, Dan knew there were nine volunteer fire departments in the county: Colvill, Grand Marais, Grand Portage, Gunflint Trail, Hovland, Lutsen, Maple Hill, Schroeder, and Tofte.

Tom Lynch and his team were also working the evacuation, and because they were well practiced and well organized, and it was early in the season, it did not take long to complete the evacuation.

At every residence those working the evacuations were also doing assessments. They were checking to see if the structure had been set up with a sprinkler. If it had, they were checking to see if there was a pump and fuel. They were also checking egress routes—means of entering and escape—to make sure they could get in and out.

They noted those structures with the sprinkler infrastructure. For these they would return and try to get the sprinklers operational. Since it was so early in the season, many of the systems had not been turned on. In fact, Mike and Sue Prom's Voyageur Canoe Outfitters had the infrastructure in place, but they—like many others in the area—thought it was simply too early to get it running. Consequently, their house, main store, bunkhouse, and several other structures were sitting in the dry, windy day in the sun, waiting for the first firebrands to appear. It wouldn't be long.

At 12:10 p.m. one of the GTVFD engines up the Trail radioed Cook County dispatch. "Fire is crossing the Gunflint Trail." The first conclusive break in the line had happened. It wouldn't be the last.

As the morning progressed, multiple loads of water had been

dropped on the flames. On the ground, wherever possible, firefighters were dousing spot fires. Now that the fire had neared and crossed the Gunflint Trail, there was another safety concern: electricity. Power lines use the highway right-of-way to carry electricity to the end of the road. The lines were held aloft by dried-out wooden telephone poles. When fire hit the poles and became hot enough, they combusted like roman candles. Once they fell, they were not only a traffic hazard, but the electrical lines they dropped could have catastrophic effects on any who encountered them, especially if their electricity was transmitted through water now being dropped over the entire area.

Someone from the GTVFD radioed Cook County dispatch. "Request CC page REA [Rural Electric Association] to shut off power—CC paged on call lineman—Schroeder called and where do they want power shut off—CC per GUNFD IC from the Seagull Station up shut the power off."

From the Seagull Guard Station/ICP to the end of the Trail, power was about to be cut.

Thankfully, telephone lines don't carry a worrisome electrical charge. Though all the firefighters had access to radios, locals still in the area and working on structure protection—such as Sue Prom at Voyageur—still relied on the phones to contact the outside world. But at 12:39 p.m. a call from a GTVFD engine near Voyageur signaled that somewhere along the Trail the line had been cut, probably by a falling telephone pole.

"Request CC ck phone at Voyageur," dispatch reported. "And if it is out report it to Century Tel—CC did and line is not working—CC 10–5 info to Century Tel and let GUNFD IC know."

Now along with the power, the phones were out to the end of the Trail.

Once the evacuation was completed, Mike Prom finally transferred his assignment as GTVFD IC to Dan Baumann, who had been shadowing Mike throughout the morning. Dan was the current chief, so the handoff was seamless. Then Mike got in his truck and drove up the Trail to help set up sprinkler systems. Once he got to the end of the Trail, he would be the structure protection liaison, helping everyone already working to get sprinkler systems up and running.

For now, Mike was too busy leading the effort to get sprinkler systems operational to tend to his own system. He had to leave Voyageur's preparation to his wife, Sue. While she was assessing what they needed, help was on the way from an unexpected quarter.

§16 MANAGING CHAOS

On the previous day Division Supervisor Tom Roach had been re-turning home from fighting a DNR fire in northwestern Minnesota when he checked in at the Minnesota Interagency Fire Center in Grand Rapids.

Tom was no stranger to fires. He started as a smoke chaser, or seasonal firefighter, for the Minnesota DNR in 1993. He worked as a smoke chaser for a couple years before joining a hotshot crew in Montana. A hotshot crew stays together six months a year and fights fires all over the country. "They go from fire to fire to fire all summer long," explained Tom. "Their fires are usually physically arduous and technically difficult. Other crews can do very technical and difficult work. But on average, hotshot crews get the tough assignments." Tom was on a hotshot crew for three years. During the fire season there are approximately a hundred twenty-person hotshot crews across the United States responding to fires.

After working on a hotshot crew, he became a smoke jumper for the U.S. Bureau of Land Management (BLM) in Alaska. Finally, after six years with the BLM, he moved to the USFS, where he was an engine captain and a division supervisor on John Stegmeir's Type II team.

On Tom's return home from fighting the northwestern Minnesota fire, he called another team member. "What's going on back home?"

"There's about a thirty-acre fire that started off the Gunflint."

Wow, that's really interesting, Tom thought. He thanked his colleague and hung up. Since he was driving by the MIFC, he stopped and walked into dispatch. As soon as the dispatcher saw him, she said, "Hey, you're a division sup, right?"

"Yeah."

"You available right now?"

"Yeah."

Within the hour he and another USFS colleague had boarded one of the air attack planes and were on their way to the Gunflint Trail.

That first afternoon and evening he scouted the area and got a feel for what was going on at Tuscarora. By the time he got to the Seagull Guard Station, it was late Saturday night. He found a spot in the station bunkhouse and tried to get some shut-eye.

Over the course of Sunday morning Tom attended the morning briefing, obtained the IAP, discussed his area of concern as the division supervisor, and went up in the Beaver to have a better look at the fire.

On the previous day Vance Hazelton had fought the fire all through that day and night. By evening he was sacked out in the back of a DNR pickup truck, watching the glow in the southwest like everyone else. And like everyone else he thought the fire was heading due west, and Sunday would be a good day.

But from early Sunday morning, when he drove his truck to Round Lake Road to have a look at the heel of the fire, he began having doubts about how this day would unfold.

Vance was now a task force leader trainee working under Division Supervisor Tom Roach. In the morning, when Vance drove down to look at where the fire was supposed to bed down for the night, he found an active fire line with trees torching. "We knew it was a pretty unique fire because of the way it was burning on that first day and how quickly it moved," said Vance. "And then in the morning when we rounded the corner and saw an active flame front and some torching going on . . . that was unexpected. It was still kind of a flanking fire at that time." But it was already beginning to change.

From the station they could see a smoke plume south, but they weren't certain about where the actual front of the fire was. The wind was shifting, and last night's right flank was turning into the fire's head, a transition it would be making throughout the day.

When Division Supervisor Roach went up in the Beaver plane to do some reconnaissance on the fire, Vance instructed his crew to help set up the ICP and to lay hose around the station. That hose, it turned out, would be badly needed later in the day.

Vance took his truck south down the road to get a feel for what they might be able to do. He was in radio contact with Tom overhead in the Beaver, and he was sharing what he was seeing on the ground. He could see the column of smoke in the south, and it appeared to be pushing north. It was not yet close enough to the station to be worrisome, but it appeared to be moving in their direction.

He looked for some safe places where he and his team might be able to anchor in and do some work, do burnouts, or put in a dozer line. But the relative humidity was low, and the wind was up, so he knew he would have to do burnouts with caution. For now, there was only so much an engine and team of four could do. But he knew two crews were on their way: the Midewin hotshots from Illinois, and a MNICS Type II crew. Both were en route, and both consisted of approximately twenty firefighters. From what Vance could see, they would be greatly needed.

When Tom Roach finally landed, he drove all the way to Tuscarora Lodge. He was just beginning to get a sense of his resources, and throughout the day he knew they would be changing.

By 9:55 a.m. he was staging bulldozers at Tuscarora Lodge. Since the fire was so new and they were still waiting for more people and equipment, Tom was working with whatever he had at hand. He knew they would be using dozers to put lines into the tail of the fire, so he asked the dozer bosses to start scouting locations. Vance Hazelton, who understood the overall plan and already had a good feel for the fire and what they were hoping to accomplish, would oversee the dozers' work.

At this point in the morning, staging the dozers meant only getting them into position so that when they finally settled on how and where to use the lines, they would be ready. Tom knew one of his primary goals was to make sure no fire burned back onto Tuscarora and its buildings. He also knew the fire had started on Ham Lake, and there was a lot of country, likely with smoldering debris, between the start of the fire and Tuscarora. The whole tail end of the fire needed to be secured.

Since this was a transition day, and since he had seen the fire was burning strong and shifting north, Tom braced himself for chaos. He had worked many fires and was familiar with the process of getting started, particularly on large blazes in bad situations with weather a contributing factor. Tom knew his day was going to be spent bringing people and resources on line.

When firefighters showed up, you couldn't just tell them to go out into the woods and fight the fire. "You have to stop and talk to them," Tom explained. "Here's our communication plan, here's the safety plan, here are some potential hazards, and here's the way you can mitigate those hazards." Tom had to bring them up to speed on not only the morning's IAP but also what had changed since then, factoring in subsequent discussions with Operations Chief Peterson and others, as well as how the fire was changing. "You have to make sure they know their assigned resources," Tom continued. "And that they know their adjoining resources. Here's the latest weather report and prediction. Here's what we've been seeing so far, and here's what we expect we're going to do. Everyone coming onto the scene needs to have the proper situational awareness, so they can safely fight the fire."

After getting the dozers set up, Tom was asked to take some law enforcement officers to the start of the fire. That occupied him for almost an hour, during which he was putting in air support requests and receiving calls from more personnel coming up the Trail. As personnel dedicated to him arrived, the calls increased in frequency.

"Hey, Roach, I'm here. Can I meet you face-to-face and get a briefing?"

"Sure," he agreed. And then he was off on another communication run, which usually ended in orders. As the day progressed he received, briefed, and assigned two Type 6 engines (one from the Colvill volunteer fire department) to tend to structures around Tuscarora, putting out any spot fires that might flare up, and reducing fuels around the buildings. He also assigned people to secure the line from Ham Lake to the resort, to make sure they doused any spot fires coming back to life.

At noon he was coordinating an air attack with Jody Leidholm, trying to douse some flames near the Kekekabic Trail. Later he received and briefed more Type 6 engines, some from volunteer fire departments down the road, including Lutsen and Maple Hill.

"I remember the start of that day there wasn't much we could do except secure the fire around the lodge and Round Lake Road. Everything else was burning from Round Lake into the forest in a northwesterly direction, so it was off the road system, and we couldn't get into it. You can't start walking cross-country through all that heavy timber and blowdown. You have to anchor, flank, and pinch when you're fighting a fire like that. You go back to the tail and make sure it doesn't take off and do some weird new thing. And then you slowly start working up your flanks. This is how you fight a fire. You work up the flanks until you get to the head. And then you try and pinch the head off."

At Round Lake Road the Gunflint Trail turns almost due north for more than two miles. The previous day the fire had burned almost due west, so even though the fire was beginning to turn and the entire right flank of the fire was becoming its head, it still had a lot of forest to burn through before it reached the road.

At the station the Trail turned due west for about one mile before swinging north again. As the day progressed, Tom figured the bulk of the resources, including the air attack and all Tom Lynch's structure protection efforts, would occur in Division B. Tom was concerned with Division A and buttoning up the tail of the fire. And that's what they did.

"We secured the perimeter around Tuscarora and on Round Lake Road. Any fire that came out to the Gunflint or any place else we could reach it, we were putting in bulldozers to try and secure the line. We were trying to keep it from creeping past the ground where it already was."

In an effort to understand his division's territory, Tom was going through the woods scouting the area, trying to get a feel for where the fire was and what it was doing and where they might be able to anchor and work. He used dozers to improve the USFS road that ran around the north-northeast side of Round Lake. They created defensible space around Tuscarora's property. There was a boat landing on the north end of Round Lake. There was an unsecured fire line on the USFS road leading to and around the landing.

The heart of this day for Tom and his team involved securing the fire's tail, trying to figure out where they could work on the fire's flank, and receiving and briefing resources so he could get them operationally engaged.

On Saturday night, by the time Kurt Schierenbeck conducted his 10:00 p.m. briefing at the Grand Marais Station, Dan Grindy had arrived. Dan had been working in fire since 1976, when he became a smoke chaser for the state of Minnesota's Carlos Avery Wildlife Management Area. He worked there for three spring seasons before moving to Oregon, where he worked on an engine crew and was a fuels technician. He returned to Minnesota, was hired as a forester for the Minnesota DNR, was a DNR district forester in St. Cloud from 1986 to 1989, then a fire program forester from 1990 to 2001, before finally becoming a Minnesota DNR area supervisor at Hinckley and then Cloquet. On this day he was one of the Type II team's division supervisors.

Since it was a transition morning, and since Greg Peterson and some of the others needed to get a feel for the fire, Dan wasn't assigned to Division B until around midmorning. In fact, the May 6 IAP did not articulate Division B, its division supervisor, or its duties. But by 10:30 a.m. Dan had a clear understanding of his duties and the severity of the task he and his team had taken on. Division B was to manage the territory from just west of the Seagull Guard Station to approximately four miles up the Gunflint Trail to where it passed Seagull Canoe Outfitters and Lakeside Cabins. While on paper it may not have seemed like extensive territory, parts of it were already threatened by a growing wildfire.

Approximately a mile west of the station the Trail started turning north. As it angled north-northwest, it approached Blankenburg Lane, the place where Tom Lynch, just moments before Dan arrived in the area, had been forced to flee. Shortly before the Blankenburg Lane turnoff, Seagull Creek crossed the Trail. Here the flames and spotting were rampant. The task for Division B was to keep those flames and spotting—if at all possible—west of the road. They did not want the fire crossing the Gunflint Trail.

Where Seagull Creek crossed the Trail, there was a large, open marsh. Visibility was relatively good, but more importantly the fuel source in this marsh area was largely dried out grasses. Since the fire was increasingly turning north, the line due west of the Gunflint Ranger Station was a flank of the fire, but due to a westerly bend in the road the fire hit the Trail with the force of a head fire. Everywhere the grass burned, and thick clouds of smoke billowed into the air, clouding north and east.

Dan and his team were set up along the Trail, dousing spot fires and trying to keep the flanking flames from breaching the blacktop.

His two Type 6 engines and the firefighters manning them spread out along this part of the Trail, and for the next sixty to ninety minutes fought the sometimes approaching, sometimes flanking flames.

"I remember the wetlands area," Dan recalled. "We had a good view across it, and I was able to see the main body of the fire quite well as it approached the road. At one point I actually saw the spotting go across the Trail and get into some jack pines on the other side." Once it got into the jack pines, the fire became so intense and the smoke so dense "we didn't have a chance to get in and do a lot. At that time, when I made contact with air attack, they informed me that the fire had already crossed the Trail with spots that I was unable to see."

That was around noon.

"Since we couldn't see down through the smoke, we couldn't work on firing pines and the surrounding area with the aircraft," Dan remembered. By this time the wind and fire were both intense, the spotting was impossible to contain, and the Trail had already been breached, so there was little else Division B could do but turn their attention to safety.

"The fire and smoke were dense across this part of the Gunflint," Dan said. "Our job was to escort people evacuating from up the Trail through this part of the fire." For the rest of the afternoon, he and his engines escorted evacuees and others trying to get down the road.

"A tree would fall across the road, and we'd clear it out," Dan recalled. "We posted people at both ends and only allowed one-way traffic through the section of the Trail breached by the fire. Once you got in there, the smoke at times was so dense sometimes you were lucky to see the pavement."

§17   THE LONG AFTERNOON

On Saturday, May 5, GTVFD Chief Dan Baumann was in Ely. He did not return home to the Golden Eagle Lodge, twenty-seven miles up the Trail from Grand Marais, until very early Sunday morning. At that time Mike Prom, who was still awake, briefed Dan on the Fire Day One activities. Mike was confident, given all the reports and what he had seen of the fire, that Sunday would be a good day.

Dan appreciated the positive news. Still, early the next morning Dan was at the Seagull Guard Station ICP in time for the 7:00 a.m. briefing, and by then he was beginning to get a sense that the tide of the fire had changed.

Because Mike was still working the fire, Dan shadowed Mike, both to get a feel for the status of the firefighting and volunteer effort, and to provide assistance and support to Mike if it was needed. Hands-on training is one of the most important aspects of learning how to fight and manage fires, which is true for all firefighting organizations. At least for the start of this day, Dan was going to let Mike run with the command, to get a feel for how it all worked.

By the time Sheriff Falk called the evacuation, Mike was ready and managed the entire evacuation in record time from the top of a garbage can lid. Dan knew from experience that when it came to fighting fires, you never got to choose where the fire would start or where you would need to position yourself to get the job done.

After the evacuation was complete, Mike headed up the Trail to assist with structure protection. Most of their work—and the work of Tom Lynch's crew and the other GTVFD personnel on the job midday—involved setting up and making sure any sprinkler systems that were available were operational. There were lots of moving parts to a sprinkler system, which is why Mike was happy to hear Michael Valentini's voice come over the radio.

Earlier Sunday morning Michael and his wife, Sally, were working on Highway 53 on a cleanup effort across Minnesota's Iron Range, near Chisholm. When Michael got the call, he worried about his house on Sag Lake Trail, which was likely in the path of the oncoming flames. He also knew that if possible, he should get to the end of the Gunflint and help out with the sprinkler systems.

Michael was not part of the GTVFD, but he had assisted with both the Cavity Lake and Alpine fires and with other efforts, particularly with regard to sprinkler system setup and maintenance. A growing part of his handyman business involved setting up and maintaining sprinkler systems up and down the Gunflint, particularly at the end of the Trail. Michael also had had a previous career as a paramedic in the Twin Cities, so he had a definite skill set to contribute.

Michael had made the trip from Chisholm to the end of the Gunflint Trail so many times he knew the quickest route. MapQuest estimates the driving time at four hours. Michael could do it in less than three hours.

By the time he approached the roadblock at Cross River, still a mile short of Round Lake Road, it was around 1:30 p.m. He could see there was a lot of activity, including a menacing column of smoke in the distance. More important, the sheriff's roadblock was erected across the Trail, preventing anyone from getting farther up the road.

At this point, Deputy Doug Rude was manning the roadblock. Deputy Rude was new to the sheriff's department, so he did not know or recognize Michael Valentini, and Michael did not know or recognize him. When the deputy stopped Michael in his vehicle, he greeted him and told him the area had been evacuated and no residents were being allowed in.

"If I can go up to the end of the Trail," Michael said, "I can help them with structure protection. A lot of those sprinkler systems are mine. And I've got parts and equipment up at my shop, near Voyageur's."

Deputy Rude knew the fire had already crossed the Gunflint Trail in parts and was threatening the entire area. But his primary mission was to keep people safe, and from what he had heard, there was a lot of structure protection already happening at the end of the road.

"From what I understand everything's pretty good," Rude said. "I think you need to stay down here, where it's safe."

Michael wasn't one to argue with law enforcement. "OK," he agreed.

As he contemplated his next move, Marilyn, who worked with Sue Prom at Voyageur and was Valentini's neighbor, came around the roadblock. She and others from up the road were evacuating supplies, important records, and whatever else they needed to from Voyageur.

Michael flagged her down. "How is it up there?" he asked.

"It's a mess," Marilyn said. What she shared about the growing chaos of the scene convinced Michael he needed to get up there.

This time he approached the deputy and asked him to get Mike Prom on the radio, to see if Prom could use Michael's help. The radio traffic over the GTVFD frequency was busy, but after a few minutes the contact was made, and Mike did not hesitate. "Definitely. Send him up."

Michael covered the distance to the station, noticing how fire and smoke were nearing the road. Judging from the way the smoke was blowing, it also appeared as though the flank was now a very long front, moving north, threatening the blacktop. When he reached the station, he looked ahead and could see fire near the road and smoke in the distance, near where the Trail turned. It appeared as though far up ahead the flames could be crowding the road.

When Michael stopped at the station, he was greeted by Dan Baumann and Don Kufahl.

"Can't go up," Dan told him. "Fire's across the road."

Overhead, tankers were dropping water up and down the line.

There was nothing to do but sit and wait. Michael discussed the situation with Dan and Don as the overhead aerial assault kept dropping loads on the fire. After fifteen minutes someone radioed from overhead and let Dan know Michael could go now; the Trail up ahead was safe. But he needed to be quick about it.

Far up ahead the road appeared covered in smoke and danger-ous, but Dan and Don waved Michael through. He put his pickup in gear and headed down the road. He had traveled about a football field when back at the station, Dan and Don were again contacted by an overhead radio.

"No, no," the aerial reporter warned. "Don't let him go. Don't let him go. It's burning again."

But it was too late. Don and Dan waved and yelled, but there was too much noise. And Michael was too far away. He did not have a radio, and besides, he was not looking in his rearview mirror; he was staring ahead. And what he saw did not look good. Still, he needed to get up the road, and the air report had said it was good, so there was nothing else to do but keep driving forward.

He had traveled this road on countless occasions but never in conditions like these. As he moved forward, he was shocked by the fire on both sides of the road. Peripheral trees and brush were flam-ing. His window was closed, but through it he could feel the ambi-ent heat, intense in the early afternoon. If he got stranded here, he would never be able to get out of his truck. If he had had time to think about it, he would have worried. But right then he was just trying to get through the tunnel of flame. He watched trees that went up in a whoosh and then continued to burn in pillars of fire.

The drive was intense. The ambient heat through his windshield and side window became so hot he could feel it on the side of his face and shoulder. While he was pushing through the area, he thought, "My god, how bad does it have to get before they say you can't go through it?" He thought that they shouldn't have let him drive in conditions like these. But they did, and he did.

Finally, after 100 or maybe 300 yards—consumed by keeping his truck away from the edges of flame, he had lost track of how much distance he had covered—he came out the other side. All he could think of was, "Wow! I made it!" He kept driving up the Trail, still stunned by his drive through the gauntlet of fire.

Farther up the Trail he crossed paths with Mike Prom, who was managing structure protection from his vehicle.

"That was crazy," Michael told him, explaining what he had just driven through.

Mike couldn't believe he had made it, but he was glad to see him. Mike gave him the address of a sprinkler system that needed attention, and Michael went off to work. After getting that system operational, he made sure Voyageur's sprinkler systems were working. The rooftops of their main buildings and most of their land were covered by sprinklers. There were one or two outbuildings that were not covered, but at this point there was nothing that could be done to protect them.

Tom Lynch, Mike Prom, and others were very glad Michael had made it through, because he began working steadily on every sprinkler system he had time to get to, repairing hoses, fixing joints, getting pumps operational, supplying propane tanks to those that needed them. A neighbor had gotten the sprinkler system on Michael's property operational, so he only visited his own property to retrieve needed supplies.

On Prom's property there was a propane refueling station, and later in the day Jody Leidholm called in the P-3 Orion to douse the area around the refueling tank with retardant. The refueling station was the lifeline to the sprinkler systems that were operational, and if it or the area around it burned, the entire station could blow, creating an explosive fireball everyone wanted to avoid.

"Valentini was running around with his head cut off," Tom Lynch recalled. At one point in the afternoon Tom and Michael were working on the sprinkler system at Seagull Canoe Outfitters and Lakeside Cabins. At another, Prom asked Michael and Steve DuChien of the Grand Marais Volunteer Fire Department to go across the Seagull River to the other side. Where the river opened up, it was called Gull Lake. There were approximately twenty cabins on that shoreline that were water access only. Though they were across the water, given the way the fire was spotting far ahead of its main front, it was very likely the fire would cross the river.

Michael and Steve went to Michael's cabin to use his speedboat to cross to the other side. When they got to Michael's dock, his boat was gone! In the middle of this catastrophe, someone had stolen his boat?!

Since they were so close to Voyageur, they walked over to use one of their boats. But they, too, were all gone!

Michael could not believe there was an opportunistic boat thief in the area, but neither did he know what was going on. And unfortunately, he did not have time to find out.

He could see that across the river Donny Olson's boat was moored to his dock. So he and Steve grabbed one of Prom's canoes and paddled across the waterway. They got in, started it, cast off the lines, and turned its prow across the river toward the cabins on Gull Lake. Suddenly the boat sputtered and died. After trying to start it back up, they checked their fuel level. Empty!

Over the next few minutes they paddled ashore to a nearby cabin and found some stored gas, which they used to top off Olson's tank. Then they continued to check on the sprinkler systems farther up the shore.

• • •

After the failed burnout near Blankenburg Lane, Fire Management Officer Tim Norman, Firing Boss Barb Thompson, and Operation Chief Peterson all returned to the ICP at the Seagull Guard Station. One of the first things Peterson did was tell John Stegmeir he needed to "get all nonessential resources out of the ICP." Eventually, by the early afternoon, IC Stegmeir decided to move his command a few miles down the road to the Gunflint Lodge. Peterson and others believed it was only a matter of time before the station was threatened or perhaps burned. The evacuation of the ICP required every capable hand pulling up tents and whatever else could be burned, which they threw into their trucks, retreating back down the road.

Throughout the rest of the morning and midday, Tim Norman and John Wytanis patrolled the area in their trucks, getting a feel for the fire and beginning to have a larger discussion with Operations Chief Peterson and others.

On the previous day the Red Book assessment tool was used to determine if the fire should be elevated from a Type III incident to a Type II incident. The assessment clearly indicated the need to elevate the fire to a Type II incident. On the previous evening Forest Supervisor Jim Sanders asked John Wytanis to sign the incident over to the Type II team, which had been done.

While on this transition day resources were pouring into the area, Tim Norman was beginning to sense, given the fire's extreme behavior and other factors, that this incident might need to be elevated to a Type I. Sooner rather than later.

First, the number of personnel arriving to fight the fire was beginning to approach an amount that indicated a much more complex and robust incident than the resources required for a Type II wildfire. Similarly, more aircraft, some of them heavy, were being called up. For now, fighting the fire by air was in large part their best recourse, given that the fire line was still for the most part away from the roads and trails they could use to get in and fight it on the ground.

Since morning, when Tim had been up in the Beaver with his colleagues, he had realized that unless they stopped this fire, it was only a matter of time before it would burn far enough north to cross the border into Canada. From the Seagull Guard Station the Canadian Border is just a few miles, as the crow flies. And through midday Tim had seen nothing to indicate they were going to stop this fire. More important, from everything he had seen, the fire was gaining momentum and—as he recalled from his morning flight—spotting far out in advance of the front lines.

In addition to the complexity of resources fighting the fire, it was almost certainly going to become an international incident. Any time an incident exceeded U.S. borders, it added to the fire complexity. In fact, by midday Tim had made a note to remind Peterson to call the liaison at the Ontario Ministry of Natural Resources and alert them to the situation. Protocol dictated that they invite an appropriate firefighting professional from Ontario to shadow their efforts at the ICP.

Then at 3:30 p.m., Tim was monitoring the radio when he heard one of their overhead aircraft report Sunday's first structure lost to the flames. Thirty minutes later four more structures were reported lost, all on Seagull Lake Road. This time the numbers were ticked off, and Tim recorded them in his notebook: "353, 323, 257 and 289." Each address was obviously someone's cabin. Presumably the first structure reported lost was also a cabin. That meant five cabins had been lost in the past thirty minutes. And Seagull Lake Road was less than a mile south of Blankenburg Lane, where they had—only hours earlier—been chased away by uncontrolled spotting flames. The solid front of the fire must have finally marched across Seagull Lake Road, destroying the unprotected structures in its path.

Again, the destruction of structures and the certainty that others were likely to be lost as the fire continued north were indicators of incident complexity. Tim, John Wytanis, and others were beginning to think it might be only a matter of time before this incident required more personnel and resources than a Type II effort could marshal.

Throughout the day Tom Lynch, Mike Prom, Michael Valentini, and many others were doing what they could to get structure sprinkler systems operational. As the day progressed, Tom and his fire engines were seeing more spotting activity at the end of the Trail. In fact, Tom was largely manning the radio, keeping everyone in the loop. Tim Norman and John Wytanis were both in the area, wanting updates on any structures that had been lost. Tom knew, from speaking with Tim and others, the main flame front was still south of the Gunflint Trail but approaching. At the end of the Trail, where he and his crew were, they were encountering heavy spotting. It was only a matter of time before the fire reached them—though that should be much later in the day or early Monday. Still, they continued to douse spot fires all around the end of the Gunflint Trail.

Tom knew they were making an impact on the spot fires. Every time they occurred they were knocking them down. He had worked his way through the Trail's End campground, dousing spot fires. But by the time he made the big circle, he needed to do it all over again.

This was the end of the Gunflint Trail. From here there were only two ways to escape. They could drive down the Gunflint Trail and—providing the Trail was not blocked by flame—get back to the ICP at Gunflint Lodge. However, from the reports Tom and others on his team were getting, they knew the Trail was in places already compromised.

They could also do the unthinkable: abandon their engines and head to the water.

Mike Prom worried about being cut off from the Trail. Earlier, as a precautionary measure, he had sent out members of his crew to gather as many boats as possible and bring them to the Saganaga

Lake public landing. If they needed to make a run for it out on the open water, he wanted to be prepared. That is why Michael Valentini had found his boat and then those at Voyageur missing. The boats were at the landing.

As the afternoon unfolded, Tom was becoming frustrated by the constant dousing of spot fires, only to have more pop up all around them. While his team was doing an excellent job knocking down the fires, everyone knew the solid front of the fire was marching toward them.

Finally, the crew on one of his engines became spooked and told him they were going to head to the safe zone, near Seagull Canoe Outfitters. The parking area there was an open space they felt they could defend if the flames approached. Also, from there, they could easily head down the Gunflint Trail if it became passable.

The concern of one of his engine crews caused Tom to have what he called a "come to Jesus meeting. The disengagement of one of my engines was a wake-up call. I still thought we had the ability to be up there, but not everyone agreed."

Then Tom got a radio call from Tim Norman, who gave him an update on what was happening south of him. The head of the fire was getting closer, and it was coming their way.

After Tom received this report, he sat down with the rest of his crew—Pete Lindgren, Mike Prom, Bob Baker, and others—and they decided there were better places to use their resources. And that is when they all steered their engines to Seagull Canoe Outfitters to make a last stand before the front of the fire blew through. Earlier, Tom and Michael Valentini hadn't been able to get all of Seagull Canoe Outfitters' sprinklers running. Without those sprinklers and with the head of the fire bearing down on them, they were certain to lose structures, if not the entire resort.

Around this time Tom took a call from Dan Baumann at the Seagull Guard Station. Since Dan had put in a call for mutual aid from Cook County's nine volunteer fire departments, seven had sent Type 6 engines to help with the evacuation and structure protection.

Now several of them were at the station with Dan, and he wanted to send them to the end of the Trail.

Tom was concerned, because he had seen the fire and smoke. He knew they could put the engines to good use, but he was concerned that with the Trail covered in places by smoke so thick you could barely see your hand in front of your face, there could easily be a head-on collision with another vehicle heading south.

Tom sent Bob Baker down the Trail to do some reconnaissance. Tom had posted volunteer firefighter Rick Johnson south of Seagull Canoe Outfitters as a kind of traffic cop. He also knew Sheriff Falk had recently come up the Trail. After checking with Falk and Johnson and then hearing the all clear from Baker, he knew there were no vehicles heading south, so he finally gave Dan the OK.

Dan took off from the station, leading a caravan of volunteer fire engines up the Trail.

18

FIRE AT THE TIP OF THE BAY

Ron Berg first visited Seagull Lake in the summer of 1973. He had paddled and camped in the area and, like many who have visited the place, was awestruck by its natural beauty. Also like many, he thought buying a cabin in the area would be like owning a piece of paradise. But as a teacher in St. Cloud, with income scarce, he believed the thought was more dream than reality.

Then in the fall of 1973 he noticed an ad in the Minneapolis Sunday paper. "Cabin for sale on Seagull Lake. $11,500, or will take a plane in trade." The address was 66 Island Road. The mere sound of it evoked a sense of adventure, solitude, and replenishment. Ron did not own a plane, but he did have a father who was willing to loan him some cash. He made an offer of $10,000 and was surprised and elated when it was accepted.

As a teacher he looked forward to the end of the school year, when he and his daughter, and then later his wife, Keli, could head north and spend the summer at his oasis on Seagull Lake. After a couple of summers at the cabin, spending time paddling and fishing on Saganaga Lake, he began to work as a summer fishing guide, which he did until he decided to move to the area permanently in 1990. During that time teachers could take a five-year leave of absence from the St. Cloud school district and still retain their tenure and seniority if they decided to return.

In the fall of 1990, when he was done guiding and not returning to his teaching job in St. Cloud, he was asked to do some cooking down the Gunflint Trail at Bearskin Lodge. He had been cooking part-time in St. Cloud at the Redwood Inn, so he was familiar with what the job entailed.

The winter of 1990–91 was one of the first times the Gunflint Lodge was open year-round. Bruce Kerfoot needed a breakfast and lunch cook for the lodge, so when he learned about Ron's availability and expertise, he offered him the job. A short time later the Gunflint Lodge's executive chef quit, and Ron was offered that job, which he accepted. He never availed himself of the opportunity to return to teaching. He became the Gunflint Lodge's executive chef for a decade, until he retired in 2000.

Over the years the Bergs transformed their one-room cabin into a permanent home.

"When we first bought it, it was a 14-by-24-foot single room," explained Ron. "We added a 20-by-24-foot addition in the late 1970s. And then we built a 20-by-24-foot screen porch sometime in the eighties. And then in the nineties we turned that into a four-season room, with a wood-burning oven coming in at one end. That's where we spend most of our time now."

Early Sunday evening on May 6, 2007, Ron and Keli Berg were returning from a weekend in Duluth when they were surprised to find a roadblock at Cross River, just eight miles from their home near the end of the Gunflint Trail. They had heard nothing about the fire and had no idea an evacuation had taken place and that their home was now threatened by an approaching wildfire.

The Cross River roadblock was at a low point on the Gunflint Trail and thick with trees. Even if Ron and Keli could see far ahead, they were too far away to see any of the fire, smoke, or commotion that was consuming the end of the Trail. When they approached, the deputy told them they could not continue up the Trail. He told them about the evacuation order issued earlier in the morning: no one was being allowed beyond this point.

"But we have cats," Keli said. The Bergs owned two cats, both at home. If there was a threat of fire, they needed to retrieve them. They also had a sprinkler system they wanted to make sure was operational. They were hoping a neighbor or firefighter had started it. Like many permanent residents in the area, the Bergs were worried that their home was being threatened by fire. It was not the first time—they had had scares in the past, most recently the Cavity Lake fire. When Ron first heard about sprinkler systems, he knew he would install one; he thought it was an excellent idea. He hoped the untested systems would remain untested, but in order for them to work they needed to be operational. He knew from experience they could be finicky. Hoses could burst, and connections and pumps could fail. The Bergs needed to get up the Trail to check on their place.

"I don't have the authority to let you pass," the deputy told them. "But if you want, you can go back to the Gunflint Lodge and talk with Sheriff Falk. Maybe he'll permit it?"

The Bergs returned down the road to the Gunflint Lodge and pleaded their case to Sheriff Falk, who had been up the Trail and knew the fire was threatening the entire area. But he had not seen it recently, so he turned to someone who had and asked, "Can they get through?"

"Yeah, they should be able to."

"OK," he told the Bergs. "The deputy will let you through. Go up to the Seagull Guard Station, and I'll have someone there escort you the rest of the way up the Trail to your cabin. Get your cats, make sure your sprinklers are working, and get out."

By the time they arrived at the Seagull Guard Station, they could see smoke and flames in the distance. There was a flank of fire burning from south of the station, across the Gunflint, angled toward the northeast. It was still to the west of the station but threatening.

Overhead, aircraft were dousing the flames, and firefighters were on the ground.

Even though the ICP had already been evacuated and moved down the road to the Gunflint Lodge, there were still plenty of firefighters around the station, fighting the flames, doing burnouts, dousing the fire, or in other ways trying to make sure the station was saved.

Here, the Bergs ran into Jim Wiinanen, who had been assigned to escort them the rest of the way up the Trail. Given the flames and smoke, the Bergs were happy to see Jim, a familiar face. For years Jim had run the Wilderness Canoe Base, almost at the end of the Gunflint Trail.

As Jim led them up the Trail, they could see fire on both sides of the road and spot fires moving north away from the road. There was fire and smoke coming from the south, and occasionally they passed a tree completely covered in flames. In one place the fire was so intense they could feel the radiant heat through their windshield and side windows. Finally, they pushed through the heaviest part of it and eventually approached the familiar Seagull Canoe Outfitters and Lakeside Cabins. Right after passing Seagull Canoe, they arrived at the turnoff onto Island Road. Their home was third down on the left.

They needed to locate their two cats, make sure their sprinkler systems were operating, and then return back down the Trail. But once they arrived, they were confronted with the unexpected.

"Reaching our house, our worst fears were realized when we saw that no one had started ours or our neighbor's wildfire sprinkler systems," remembered Ron. "Keli and I tried to round up our cats. We found one, but we couldn't find the other anywhere. We started frantically grabbing items and hauling them to the car. Still no sign of our other cat."

Fortunately, from his years managing the Wilderness Canoe Base, Jim Wiinanen was familiar with sprinkler systems and began to assist.

"I yelled to Jim to please help me get ours going," Berg said. "As we ran down to the lake, I saw that at Lehigh's, my neighbor to the north, a ground fire was burning near the lake, right up to our shared property line. We frantically put the intake into the water, primed the pump, turned on the propane, and started the pump. Water was coming out of the sprinkler heads. Jim then hiked the short distance through the woods and started the Rands' pump, our neighbor to the south. Soon I heard the welcome sound of their pump running. When Jim returned, we hiked gingerly through the ground fire to Lehigh's pump and tried to start it, but could not get it started."

Meanwhile the fires they had driven through to get there were continuing their steady march to the northwest, pushing spot fires out in front of the main flank, one having landed at Lehigh's. Those same spot fires were threatening Seagull Canoe Outfitters at the tip of the bay, as well as the Bergs' place and all the others down Island Road and farther beyond to the northwest.

"Keli was carrying buckets of water from the lake, trying to extinguish the ground fire at the neighbors. I found the keys to the snowmobile and moved it away from the slowly spreading fire."

The Bergs' place is on the western shoreline of the Seagull Lake bay. There are approximately fifty yards of water separating their shore from the eastern side of the bay. Along the eastern shoreline there were five cabins, so close to the Gunflint Trail they had open turnouts rather than driveways. Unfortunately, the fire was coming straight up the Trail, and those cabins were in its path.

"Soon after starting our pump, it lost its prime," continued Ron. "Jim surged the intake pipe back and forth in the water, and it finally started pumping again."

Meanwhile, Keli was still fighting the fire at Lehigh's.

As the Bergs and Jim fought the flames, the fire finally reached and began attacking the opposite shoreline.

"In what seemed like mere minutes the opposite shore was a wall of flames for as far as we could see in either direction. There was tremendous roaring in the air from the fire. . . . A thunderous explosion startled us as a propane tank exploded from the intense heat of the fire. Several more explosions followed as more tanks blew up. We were too busy to notice that just up the hill from us, the fire had crossed the road, cutting off any escape by car for the time being."

They made sure their cabin and their neighbor to the south still had operating sprinkler systems. Both continued pulsing a steady stream of water. But they had been frustrated trying to start the pump for their neighbor to the north. They had, however, finally managed to get the ground fire doused.

Finally, the Bergs found their second cat.

"Jim saw the fire cross Island Road and left to see if there was a safe place for us to go down the road if we had to," Ron said.

"It was getting darker," Jim added. "And at that point propane tanks were blowing up across the bay. That was a sound I will never forget."

Jim saw the flames up the road had burned down enough to enable them to leave. He returned, thinking the Bergs now had their sprinkler and one of the neighbors' operating, and they had located their second cat.

"OK," Jim said. "It's safe to get out of here."

But there was still too much work to do. They had not yet been able to start Lehigh's system. They could see spot fires in the distance, and even though they knew their sprinkler system was operating, they had seen enough of this fire to know that sometimes the surest intervention was human intervention and that if they wanted to ensure their cabin and the neighbors' property remained safe, they needed to remain and do what they could as the fire continued coming.

"We're going to stay," they told Jim.

Jim, like the sheriff and his deputies and Mike Prom, who led the evacuation, knew the Bergs could not be forced to leave. Besides, he was friends with the Bergs, and he understood.

"It took me twenty minutes trying to decide whether to stay and be of no use or get out and try to help somebody else," Jim said. "I finally decided to leave, and that's when I had to drive through fire to get to the Gunflint Trail."

When Jim drove away from the Bergs' place, there was so much fire down and across the opening to Island Road he was focused on driving through the flames. The smoke and fire also obscured his view of the end of the bay, where Seagull Canoe Outfitters and Lakeside Cabins was located. But when he rounded the corner and started heading down the Trail, what he saw in the open parking area around the outfitters' buildings was both startling and a welcome surprise. There were several Type 6 fire engines, and the crews who manned them were using all the hoses and water they could gather to fight the threatening flames.

All day Jody Leidholm and his air attack team had been fighting the fire from the rear, along its flanks, and at its front. In addition to being called in for structure support and to help at various locations, the air attack had largely worked along the north side or right flank of the fire, trying to slow it down or keep it from crossing the Gunflint Trail. Earlier in the day they could not stop it, and from the trajectory of the fire Jody worried the Seagull Guard Station was in trouble.

Sometime after 4:00 p.m. Jody, who had a better vantage point to view a large segment of the fire from above, called Greg Peterson. "We're gonna have a problem with the guard station, and we better spool up and make a stand to save it, because it's threatened."

On the ground Vance Hazelton and his crew and others were performing test burnouts around the helipad across from the station, as well as on the open ground west of the station.

"We were trying to do a little burnout there to try and secure that area," explained Vance. "But every time we put down fire, the conditions were just so snappy that it was taking off and getting into some small crowns, so we had to go really slow. We didn't want to put down more fire than we could hold on our own."

The flank or front of the fire—at this time it was difficult to distinguish which—was spread in a burning line from the southeast, still south of the station and helipad, to the northwest. But because the wind was blowing north, it was only a matter of time before the station and its structures were directly in the fire's path. The air attack crew began dropping water and retardant on the fire line, trying to suppress its steady northward progress.

By 4:15 p.m. Division Supervisor Tom Roach was at the station with Greg Peterson. Division B Supervisor Dan Grindy was also now on the scene. The station was being threatened, and there was an all-out effort to save it.

Earlier Tom Roach had sent his dozer crews to the helipad to clear space around the area and to clear brush, grass, and anything else that could burn from around the fuel tank near the base. They were trying to make sure the radiant heat from the oncoming flames, which appeared certain to breach the area (it was only a matter of time), did not grow so hot it would cause the fuel tank to explode.

There are power boxes in the ditch all along the Gunflint Trail. While the power had already been cut, if the power boxes burned, it would slow the return of power to the entire area. To save the power boxes, Tom also had his crew make sure each of the units was free from surrounding fuel.

At 4:40 p.m. the MNICS Type II crew arrived. The crew numbered twenty firefighters. Normally, Tom would spend some time bringing strike team leader Jeremy Bennett, the head of the crew, up to speed. But because the situation was growing more dangerous by the minute, he immediately dispersed the MNICS crew across the road to the helipad to help secure it.

The air attack continued dropping water on the fire flank, trying to slow or stop its advance. Larch Creek runs due west of the station, and sometime in the late afternoon the fire began approaching it. The angle of the fire line was still to the southeast, and it was still pushing to the north-northwest, but the wind was still turning more north. That northern direction meant the flanking fire was starting to look more like a fire front. And the part of the fire front that was now only fifty yards south of the helipad was beginning to come on.

When a fire is spotting as badly as this one, it is almost impossible to determine fire front from flank from new bloom of flame far out ahead or to the side of the front.

"As soon as we put down fire, any ember that would cross the line was taken, and we had 100 percent POI [point of ignition]," explained Vance Hazelton.

Some of those embers would bloom just a few yards away. But the wind was gusting up to 30 miles per hour, so other embers would be picked up and carried more than a quarter mile before they landed and bloomed.

Jody needed to get a feel for the fire's flank, front, spotting, and direction, so he directed his pilot to fly due west of the station.

When there are layers of smoke between the ground and at all altitudes in the air, it is important to maintain good visibility. If you were to fly blind, you could navigate via your instruments, but in the chaos of fighting the fire, you could also run into something, especially a plane or helicopter.

Near the northwest the smoke plume was growing so large it rose high and wide into the air. Just west of the station the road goes due west before it turns almost due north, a mile or two down the road. All afternoon as the fire marched, Jody and Tanker 21 had been trying to spread a line of retardant along the south and then west side of the Gunflint, trying to keep the fire to the south and west of the road. They were trying to trap it and prevent its progress north, toward an area of the Trail well populated with people and structures.

To get a clear sense of where the fire head was, the pilot could fly high up above the smoke plume or try to get under it. From that vantage point, Jody might be able to determine the fire's head, speed, and direction and what other structures might be in danger.

"I asked some of the Tanker pilots to see what they could see up there," Jody remembered. "Because they have heavy airplanes. And none of them said they could and nor would they. It was too smoky, turbulent, and dangerous."

Then Jody radioed John Bell, the chief pilot at the time, who was also near the top of the huge and growing smoke plume. Jody asked John if he would go in and do some recon.

"I'll see what I can do," John radioed back.

"I watched him pull his nose onto the right shoulder where the flank turned into the head, and he poked his nose in there," Jody said. "Then I watched him make a hard right turn, and he called me back."

"Jody, I can't get in there," John said. "No way. The wind is so strong, and the column is rising because of the major crown fire. But it's also blowing horizontally."

"Even for a 35,000-pound airplane," Jody explained. "He wasn't going in there."

There was so much smoke in the air that Jody, now west of the station, couldn't tell flank from front.

"We were trying to get under the column," Jody recalled. "Now and then the column would lift. We wanted to get a better look at the fire below."

They were approaching the eastern shore of Seagull Lake. The column appeared to lift for a moment, and the pilot thought he could get under it. They saw structures, the shoreline, and Seagull Lake, and they started to turn a corner.

What they could not see was that this fire was burning so intensely, there were actually two heads to it—both contributing to the huge smoke plume. As soon as they turned, the largest of the two columns appeared directly in front of them, like a thick smoke wall. The wall came up so suddenly the pilot did not have time to turn or dive, and suddenly they passed through its edge.

They were blind. The only thing they could see was smoke. Inside the cockpit they watched their instruments. The altimeter was telling the story of a wild ride up the smoke column. They were caught in the smoke-filled updraft of wind and hot air. In less than a minute their Beechcraft rose one thousand feet. It is never a good idea to go into a major smoke column, for fear of losing an engine or becoming inverted and losing complete control of the aircraft.

Finally they rose high enough so the column began to thin. As soon as they hit the less turbulent air, Jody's cell phone flew out of his pocket, and everything in the plane was tossed about. Due to the sudden, extreme acceleration and then deceleration, everything in the plane was for a moment suspended, weightless. It was chaos, but once the pilot regained control of his aircraft, they moved off into clearer air, thankful (and frankly lucky) the path of their blind, wild ride was void of anything that could have brought them both to a sudden explosive end.

19 LAST STAND

Deb Mark and Dave Truehart, Seagull Canoe Outfitters' proprietors, had already evacuated to Grand Marais along with their staff. It had been less than a year since the Cavity Lake fire, which had also impacted the area and their business. Deb recalled that during that fire she was evacuated and worked for a week out of her Suburban. Earlier in the day, when the evacuation was called, Deb, Dave, and their staff knew what needed to be done, though being familiar with the process did not make it any easier to digest. They worked quickly to gather up their records and whatever else they could from their store and office, loading the Suburban. Dave and a coworker had been trying to get the sprinkler system up and running; however, the previous November the power company had inadvertently severed their sprinkler line, and they had trouble fixing it.

By the time Deb, Dave, and their staff left and headed to Grand Marais, they did not feel good about their business and what the day—and night—was going to bring. In fact, given the weather forecast and comments about the fire from everyone fighting it, Deb wondered if there would be anything left when she returned.

Seagull Canoe Outfitters was approximately a half mile north of Blankenburg Lane and about a mile east of Trail's End Campground. In some ways it was the gateway to the end of the Gunflint Trail.

From the outfitters, the Trail continues due north for about a quarter mile before bending west and then southwest and then turning northwest to its circular conclusion at Trail's End Campground, a two-mile meandering drive. The entire area was now in the path of the oncoming fire, which was why spot fires had increasingly been covering the region. When Tom Lynch arrived at Seagull Canoe Outfitters, judging from the smoke and spotting, he felt certain the front of the fire was not far off.

For most of the afternoon, Dan Grindy and his Division B team had been managing escorts, keeping the Trail clear, and working at the Seagull Guard Station. Now, as day was changing over into night, he and his team steered toward Seagull Canoe Outfitters to assist.

The entire area south of the outfitters was on fire. In places spot fires had blossomed into something much larger and then connected up with other spot fires. There was no precise fire front, but more of a contiguous region of fire with some parts more intense than others, all of it moving in a wide fan north.

Approximately three-quarters of a mile south of Seagull Canoe Outfitters, Seagull Creek Fishing Camp was on the right side of the Trail. The camp had a main house and several cabins and had been in the exact center of the flames as they traveled north. Cabins along Seagull Lake Access Road and Seagull Lake Road, two turnoffs just a short distance south of the camp, had also been in the line of the fire. Many or all of the cabins along these roads had already burned.

As Dan Grindy drove beside the camp, he was stunned by what he saw. "It was late in the day and the sun was almost down, and the entire scene, with all the smoke, looked black and white. It looked like the house and cabins had been covered with snow," recalled Dan.

For some of the water drops, the CL-215s mixed a mild detergent with the water. Once the soapy water dropped from the plane, it aerated as it fell to the ground, turning into suds. The suds have a broader painting area and hold moisture longer than water alone.

Soapy water also penetrates wood better. Unlike most of the other structures in the area, the camp's had survived—and continued to survive—the pulse of fire moving through the area, because of the suds.

As Dan neared Seagull Canoe Outfitters, he recalled the chaos of the scene. "I remember driving up the Gunflint, and there was fire running through the crowns, and some propane tanks started to blow. Luckily, Seagull Canoe had a big parking lot beside it." This was a clear area where Dan, Tom Lynch, Dan Baumann, and others with their trucks could all set up and begin to work.

"Fire was coming through the crowns up on a hill," Dan continued. "I remember as I was driving up, and in my mind I thought that the house across from the Seagull Canoe Outfitters was a goner as the crown fire approached. There was a structure team up there, and he [Lynch] got them set up, and they were able to knock the fire down in front of the house up on the hill. The fire spotted on the back side and started to run up the hill again toward the house." But the team was able to knock it down, too. "That parking lot gave us room to stay there and work and not be in so much danger."

Dan Baumann had led his group of volunteer engines through the front lines of the fire. They had arrived at Seagull Canoe Outfitters at approximately the same time as Lynch, Grindy, and their crews. By now there were approximately eight vehicles at the outfitters, several of them Type 6 engines. The wind was picking up firebrands from the front lines and spreading them everywhere. In a very real sense, fire was falling out of the sky.

Tom began positioning more trucks and men throughout Seagull Canoe Outfitters and its buildings. Fortunately, the outfitters was at the tip of a long Seagull Lake bay, so there was an entire lake full of water nearby. Also, the main store and office structure of the resort, as well as the cabins and the nearby boathouse, had a lot of open space around them. The boathouse sat very close to the road and was the first structure one reached when driving north up the Gunflint Trail. There were brush and trees due south of the boathouse, but otherwise there wasn't a lot of nearby fuel.

The fire was approaching from the south. Engines and men set up around the buildings and started spreading hoses, filling their tanks with water, feeding nozzles into the lake, and putting out spot fires wherever they occurred.

As the sun, already muted by the amount of smoke coming from the south, settled into the western horizon, a major battle unfolded at Seagull Canoe Outfitters. Not long after they arrived, the front of the flame hit them. Hoses were spread across the area like a latticework of heavy, flexible pipes, every one of them thick with water. Bob Baker was stationed atop the eastern bunkhouse, shooting a steady phalanx of water at the juggernaut of flames. As the fire roared toward them, their efforts began to bend the fire so that it passed around and over them. Everywhere any spot fire dropped, it was snuffed by the massive amount of water now pulsing over the property and buildings.

Many of the firefighters were virtual rookies. Some of them, Tom noticed, when he looked around during the height of the fight, were standing tall and holding the heavy hoses. One of the skills seasoned firefighters learn is to crouch and hold a hose; they are so heavy that standing fully erect forces you to hold more weight, making you tire sooner. Some of the engines were brand-new, and these, too, got a literal trial by fire.

Jim Wiinanen was manning a tanker truck. Whenever one of the engines needed water, he used the tanker truck to refill them. One of the engines that joined the fight later in the day was from the Leech Lake Fire Department. All afternoon Tom had been amazed by what this small engine had done. And now, as the fire moved up the Trail, the Leech Lake engine went back to Jim's tanker truck, refilled, and departed to fight more fire. Through this last stand, the Leech Lake engine made several return trips to refill its tanks, driving out to reengage with the flames.

Occasionally, as the evening darkened, Jim peered down and across the bay toward the Bergs' home. He was comforted by seeing headlamps in the distance, through the smoke and gloaming, knowing they were still conducting their own fight. He was also comforted by not seeing as much fire down that side of the bay as on the other side and at the outfitters.

As the evening progressed, fireballs pushed forward and pounded the firefighters. The fighters repelled them by pulsing a steady stream of water against the onslaught of flames. Finally, the fire front pushed passed. There was still much fire, but with the dark, the air calmed—though only a little. The fire slowed but continued its steady march to the north-northwest.

Not long after the smoky sun settled into the horizon, Tim Norman, John Wytanis, Greg Peterson, Carlene Youkum, and Forest Supervisor Jim Sanders arrived on the scene. They were shocked by the crazy chaos of it. Engines were askew, thousands of feet of fire hose were strung every which way. The place looked like a weaver's nightmare with hose spread across the area in a discombobulated mat of fattened threads.

The energy and stress levels were palpable in the smoky air. The flame had hit them hard, and they had had no choice but to be reactive rather than proactive. Looking around, Tom Lynch could not believe that in the mayhem no one and no buildings were lost. The place looked like chaos, and men were still spraying down buildings and spot fires. It had been a truly Herculean effort to knock it all down and bend the fire to the water's will.

"Spot fires were everywhere," recalled Tom. "Firefighters were on the tops of buildings. Bob Baker was up top on the east side of the Trail, standing on a bunkhouse roof. We had quite a few people down on the west side, trying to save the Kevlar canoes. The radiant heat was extreme, but we were able to corral the fire around Seagull Outfitter's buildings. We were able to steer the fire around. Everyone was fighting fire. Whoever was up at the end of the Trail was in that parking lot at Seagull Canoe."

Usually, reflected Tom, you did not see the immediate impact of your efforts. But on this evening the evidence of an extreme firefight and its aftermath permeated the entire scene, like the ground of one battle in a bigger war that was fought to the brink of failure but ultimately had been won.

Forest Supervisor Jim Sanders was the head of the entire Minnesota Superior National Forest. By most accounts, Sanders was not the kind of CEO who managed by sitting in his office. In fact, this was the first day of many he was about to spend patrolling the Ham Lake fire. Ultimately wildfires like these fell under his command. He wanted to see them, and the people who were fighting them, up close.

Even though Tom Lynch was from the DNR, Jim knew Tom from other firefights on which Tom had worked and Jim had been present. In fact, Tom had been working on fires so long, he knew all of the senior USFS personnel on the fire that night. When Tom saw Jim and the others approach, he was pretty sure they were going to want an update on the firefight, a tactical briefing. But clearly, Jim recognized what had happened here, because his demeanor conveyed nothing but respect, and the first question out of his mouth was, "How are you doing?"

It was one of those moments that struck Tom, because it was so unexpected. On this evening, in the middle of this fire, Forest Supervisor Jim Sanders must have had a lot on his mind. Technically, Jim would have to OK the elevation of the firefight from a Type II to a Type I. The gravity of that decision could not be overstated. Depending on whether it was the right decision, it could have a profound impact on a person's career. If it was a seriously wrong decision, it could spell the difference between early retirement and staying on.

"It's been a long fight," Tom said. "But thanks for asking."

They enjoyed some small talk, with the fire and water all around them. The feeling Tom got from Jim was similar to a son–father connection—genuine caring, rather than a forest supervisor to a person a few rungs down the ladder and from a different organization. And for that, Tom was thankful. He sensed Jim's calm, caring demeanor, and it had a profound impact on him and the other firefighters assembled at the scene. Finally they began the tactical briefing, and Tom shared his perspective and what he and his team had been doing on the fire since early that morning.

Tim Norman and John Wytanis were also discussing what they had seen of the fire. The sentiment of the group was that everyone at Seagull Canoe Outfitters should disengage and head back down the Trail. It was not only late, but the rest of the land north-northwest of this last stand was burning, or soon would be. There were spots that had been or would be saved. Every location with an operational sprinkler system was holding its own against the flames.

But the night was only just beginning.

Like a prizefighter who had just won a big round, Tom did not want to give up the fight. With the adrenalin of their most recent fight still coursing through his veins, Tom wanted to keep on, to take his engines up into the fire, and continue defending territory and saving homes. But he knew fighting fires in the dark was dangerous business. And besides, he had nothing but respect for Sanders, Peterson, Norman, and Wytanis. It was their general opinion that the best course, for now, was to disengage and send these fighters home until morning. They had fought all day. They needed rest.

Tom respected their decision and told his troops to reel up their hoses and drive down the Trail. Tomorrow was another day. For now, they needed to step back from the battle, if only for a few hours.

There was, too, the work and rest timeline. Although several of these firefighters would have liked to stay on and fight with Tom, regardless of personal safety, they had—like Tom—been working on the fire since dawn. USFS and DNR regulations regarding fires dictated that no one should work more than sixteen hours in a twenty-four-hour period. Firefighters were required to take at least an eight-hour break. There are good reasons for the rule. Tired people make stupid decisions that can lead to costly errors, injury, and sometimes death. Many of the workers at the scene were starting to bump up against the sixteen-hour maximum, so all things considered, it made sense to disengage.

While everyone was rolling up their gear and getting their engines ready to head to their stations, Sheriff Falk approached Tom and wondered if they should keep one engine back. There were the Bergs and a handful of other civilians (home and cabin owners who would become known as the Seagull Seven) down Island Road, just around the corner. These seven civilians who had decided to stay through the evacuation, or had made their way through the sheriff's barricades, had been fighting the flames and would continue fighting.

Dan Baumann and Rick Johnson were nearby, and Dan volunteered to remain behind with his engine. Rick joined him, and together they made sure they had enough water to head off and start fighting fires and doing what they could.

After his engines had all reeled up their hoses and begun heading down the Gunflint Trail to their stations, Tom headed out with them. He heard there was some action down at the Seagull Guard Station, and with the adrenalin still kicking through his veins, he was curious to go have a look.

Tim Norman, John Wytanis, and Jim Sanders remained near the end of the Trail and began having a serious discussion about elevating this fire from a Type II incident to a Type I. Tim began reviewing the various sections of the Redbook that distinguish the two responses to a wildfire. When they considered the fire behavior (extreme) and the increasing resources required to fight it, along with other factors, including organizational complexity, logistical support, personnel, and the complex air operations involving multiple aircraft, Tim and John were increasingly convinced the incident category for fighting the Ham Lake fire needed to be elevated. The demands of the fire, considered with lots of other factors—for example, it was only a matter of time before this fire spread into Canada, when it would become an international incident—indicated their effort to combat it with a Type II incidents approach was simply insufficient and would only be more so as the fire grew.

While Forest Supervisor Sanders listened patiently and considered the counsel of his two fire experts, he hesitated making an immediate determination. The previous year the Cavity Lake fire had been elevated from a Type II to a Type I, and while it was the right call to make, perhaps Jim recalled how much that fire cost. While the expense of the firefight was no reason to refuse to elevate it, no one could deny it was a consideration—especially for the senior leader who would ultimately have to make the decision.

All day Voyageur Canoe Outfitters co-owner Sue Prom had been working on numerous fire fronts. Earlier in the day she had gone door-to-door with another firefighter, checking off her evacuation lists. She made sure her own Voyageur's team was evacuated.

"Things were going from bad to worse at the end of the Trail," she reported in her blog post regarding this day. "Our power was lost, our phone lines were down and we were right in the line of the fire. . . . I went back to Voyageur to evacuate Marilyn, Ian and Theresa. I told them to load some belongings, the dog, my photo albums and some of the children's favorite things. I wanted them out of danger and ready to go as soon as the opportunity to travel back down the Trail arose."

She transported her kids down the road to her neighbor at Gunflint Pines Resort. Later she worried about Voyageur and getting their sprinkler system up and running. She had been working to put out spot fires all over her property and others' up and down Sag Lake Trail and to the end of the Gunflint.

"The wind was strong and relentless," she reported. "We knew the fire was going to consume the end of the Trail."

Because Sue was so caught up in the fight, she did not evacuate with her coworkers at Voyageur. "I wasn't with the rest of the group because my heart did not want to leave the end of the Trail. My mind was telling me to go to the kids, but it was not an easy decision to make."

As a volunteer firefighter, she had been on the front lines, behind the front lines, and over them so many times during this fateful day, she was exhausted. "By the time I headed down the Trail, flames were jumping every which way. Tree limbs were falling from the sky, ashes were raining down, and fire consumed both sides of the Gunflint Trail."

Sue was behind the wheel of her vehicle, heading into a tunnel of flame. She could not believe what she was seeing in front of her and to either side. The smoke was obscuring the road's shoulders, and the fury of flames was closing in on either side. Trees were torching, and suddenly she realized that if one of them fell and she came up to a flaming, fallen tree blocking the road, her only recourse would be to back up. And what if, behind her, one of those flaming torches she had now passed by also fell, blocking her retreat?

She could feel the radiant heat through her windshield and side window. It was so intense she leaned away from it, holding onto the wheel, trying to keep her vehicle centered through the tunnel of flame and smoke, worried her tires might melt off their rims, worried the blacktop of the Gunflint might melt.

"Just when I thought I couldn't stand it any longer the smoke cleared and the flames subsided. I had made it and I had made it alive. While I was relieved for my own safety I prayed for all of the others left at the end of the Gunflint Trail. I didn't know if they would make it out and I didn't know if there would be anything left when I returned."

When she finally reached her children that night at Gunflint Pines, somehow she found time to open up her blog and write:

> No lives have been lost and for that we are thankful. Fire crews
> showed up from Hovland, Grand Portage, Lutsen, Colvill, Maple
> Hill and Grand Marais Volunteer Fire Departments to risk their

lives to help save the lives of others. That they did. They worked cooperatively with the Gunflint Trail Volunteer Fire Department to evacuate residents from the end of the Gunflint Trail and to start sprinkler systems at as many cabins and businesses as was humanly possible. These people are all heroes. They deserve to be honored for their heroic efforts.

Today was the longest nightmare I have ever experienced. The wind blew the flames north and wouldn't quit. The flames ate everything in their way including buildings and all the trees in their path. People scurried to get out of the path of the fire and to make their break down the Gunflint Trail to safety.

Tonight we will wait. We have no idea if we will have a lodge or home to go back to. But we are all safe and sound for tonight. Tomorrow will be another fight, a fight with Mother Nature. Please pray.

THE PALISADES, SEAGULL LAKE, EVENING—Near midnight Layne and Lee donned their headlamps but did not turn them on. The night skies to the east and south were so filled with fire, the evening was lit up like the dragon-fired landscape in a postapocalyptic movie.

They climbed again to their overhead viewing place. They were now perhaps a mile and a half west of the edge of moving fire. The distant flames looked like a blood-red dawn with different parts of it dark and light. Some of the light was intense, glowing bursts that signaled either structures or thick tree stands or gas tanks going up. The fire was making a run up the distant peninsula, and they knew there were resorts and cabins in its path.

"The light from the fire was so intense that it cast shadows," recalled Gus. "It was like the most spectacular aurora borealis display ever, awe-inspiring yet deeply unsettling. The fire's light was so bright at midnight that we could walk around without our headlamps turned on."

Near the entry to the peninsula was Seagull Canoe Outfitters. A little farther up the road was Voyageur Canoe Outfitters. In the distance they recognized the area where Trail's End Campground and the Wilderness Canoe Base were located, now clearly in the path of the flames, in the middle of the conflagration. The Wilderness Canoe Base had a lodge and chapel. If you turned up Sag Lake Trail, the only road to bisect the peninsula, there were at least twenty cabins, including Bob Monehan's, all of them now threatened.

But they could see from their eagle's aerie that the flames were at least a mile or more east of them. From this vantage, the plume of smoke and fire rising out of the near horizon was blowing almost due north. For the past day and a half, they had been close enough to the wildfire's start to be periodically immersed in ash and smoke, sometimes thick, at other times more tolerable. And the wind had not abated, at least not enough to allow them safe passage to their cars. Now, judging from what they had seen of the fire's path, they assumed their vehicles had succumbed to the flames. Layne had parked at the last parking area at the end of the Gunflint, at the end of the area the fire was now consuming. Gus and Lee had left their vehicles at Voyageur. It would be some kind of miracle if any of them survived.

Normally, in the evening and through the night, the region's loons were calling. But another sign of the night's abnormality was the silence. There were no loon calls piercing the preternatural light of this dark night. They hoped whatever deer, moose, bears, foxes, and other animals that were in the path of the flames had enough foresight and good luck to escape the deadly yellow flower.

Safely west of the fire, they could watch this major natural event unfold. They hoped everyone up the peninsula had time to evacuate. They hoped the structures had operational sprinkler systems and that they worked.

They were not going to be able to write the article they had anticipated. But they were eyewitnesses to a historic fire that was already consuming structures and supplies, not to mention acres of forest. They suspected when it all blew over and continued on its path into Canada, and they got out—if they got out—they would have material for different articles, depending.

But now they were thinking more about getting out alive. For now, they just focused on keeping themselves safe.

All through the night Lee and Gus took turns unzipping their tent fly, making sure some sudden change in the fire's path hadn't turned the voracious flames on them. All of them could not help making plans for an escape in the event they needed to make a run for it. The planning and periodic checking made for a long and worrisome night.

FIRE DAY THREE

20

SAVING THE SEAGULL GUARD STATION

Since the previous afternoon's evacuation of the ICP, the frontline march of the fire had been unrelenting. Jody Leidholm's air attack and the efforts by firefighters on the ground did much to slow that march. But persistent winds, low relative humidity, high temperatures, and 100 percent probability of ignition—which stunted or prevented the use of burnouts—all contributed to the inevitability of structures being lost.

West of the Seagull Guard Station, the air attack and ground resources struggled to keep the fire south of the Gunflint Trail, but it finally breached that line a little after Sunday noon. Until 3:30 p.m. Sunday the only structure lost to the flames was an outbuilding at Tuscarora. Every other Tuscarora structure was saved, and even as the fire continued to consume acres of forest to the west, it was 4:00 p.m. Sunday before the next few structures were lost: four cabins along Seagull Lake Road. As the northern right flank of the fire pushed west and north, with spotting far out in front of it, more structures were lost. Before Ron and Keli Berg, Jim Wiinanen, Tom Lynch, and others witnessed the burning of several cabins to the southeast of Seagull Canoe Outfitters, several other structures had been lost. But the firefight had been so intense, so widespread, and still so understaffed, there had not yet been any effort to ascertain the full extent of the damage. People were still engaged in the fight and were still hoping to save more structures.

The fire's march continued, and many cabins, homes, outbuildings, and more were directly in its path. Among those buildings were several structures at the Seagull Guard Station. By late Sunday evening much had been done to save this important group of structures. CL-215 tankers and the P-3 Orion had all dropped numerous loads on the fire and in front of it, trying to paint a path over which the fire would not jump. But the wind and spotting even after dark continued, well beyond the line they had hoped would not be breached. Teams of firefighters spread out around the helipad and fueling station, across the Gunflint Trail from the station, continuing to douse spots and prepare for the possibility of conducting burnout operations.

"We were just trying to keep it in check," explained Tom Roach. "That involved using the fire engines to douse spots and flare-ups. We wanted to keep the fire on the ground, but it kept trying to get up into the crowns of the trees."

Vance Hazelton remembered the wind just before dark and the consequent futility of his team's burnout efforts. "That's not the ideal time to do a burnout, especially if you have switching winds. You're just adding to the fire. We were just trying to hold what we had. We had to wait until the relative humidity came up and things mellowed out a little."

Vance, Tom Roach, Dan Grindy, Greg Peterson, and almost everyone else thought they would have more of a fighting chance after dark. Fires are difficult to fight in the dark, but if the Ham Lake fire behaved like most, they could get some serious work done once the sun had completely set.

"So we were waiting," recalled Tom. "If we put anything on the ground, it could have been immediately explosive, and we could lose the fire right here where we were at."

The station marked the northern end of Tom's geography. And since the firefighters felt as though they had done almost everything they could for the area farther west and north of the Gunflint Trail, most of Peterson's resources, including Division Supervisors Roach and Grindy, were continuing the fight to save the station.

"We had to be very cautious about firing at the appropriate time and place," continued Tom. "The winds were howling. They were adverse to our line. They were quartering our line if not straight on. We were waiting for the humidity to start going up. So we didn't start any hand ignition until closer to dark."

The Midewin hotshot crew of twenty firefighters had been working farther west. At 8:40 Sunday evening they returned to help Tom with his fight to save the station.

"I gave them a briefing at 8:40. We started firing at 9:10 p.m. My notes at 9:20 say 'very active fire, manageable.' I had forty-one personnel assigned on the burnout operation. At 9:30 the relative humidity was still 30 percent. I had 10 to 15 miles per hour with gusts to 30."

Over the next hour his team spread out around the fire perimeter, dousing spots, fighting to keep the flames from crossing the Gunflint Trail, which would threaten the station from their own burnout. Tom's team was spread out all over the area south-southwest of the station, across the Trail, and it was pitch black. Firefighters were wearing headlamps and fighting flames, but there was a lot of smoke and ash, and the situation was chaotic. Work was getting done, but fighters were just trying to hold what they had in their burnout operation under control.

At 10:40 p.m. Tom noted they were getting "heavy spots" all over the place.

"You don't want a scattered quail approach to fighting fires," he explained. "It was starting to get to this real chaotic phase."

Then the unimaginable happened. Tom lost his line. His fighters were spread out across the wide area across the Trail, southwest of the station. And suddenly there was a crown fire push from the southwest—a strong burst of flame.

"There was a heavy push from the main fire," Tom recalled. "The main fire just came bearing down on us and overwhelmed what we were trying to do."

At this point Vance Hazelton and his team of four were in the field with everyone else, in radio contact with Tom. Vance and his small team—like the rest of the fighters—were putting out whatever fire spots they could get to. But they could not get to all of them.

"Burnout operations had been called off because we were getting spread thin and were unable to hold the line," explained Vance. "We eventually started to get multiple spot fires within the immediate footprint of the guard station compound. At one point I believe I counted about ten to fifteen spot fires near the compound. We had people running everywhere attempting to catch the spots in the darkness, but more just kept popping up."

Tom called Vance and asked him how many spot fires he was seeing.

"Ten to fifteen," Vance said.

"How many are you handling?"

"About four."

At that point Division Supervisor Roach called all his resources back to the safety zone. At 11:10 p.m. Roach began confirming all his resources were in the safety zone by the station. Everyone began filling their engines and assessing the situation. Tom was calling off each resource head and asking them to indicate their presence. He went down the list, and by 11:15 p.m. every person, engine, dozer, and all other resources were accounted for.

"He had a plan in mind," recalled Vance. "But he wanted to get accountability on everyone because everyone was so spread out and we had spot fires and equipment running everywhere."

At that point Tom dispersed the engines and hotshot squads around the compound in a more orchestrated fashion. He spread everyone out into strategic fighting positions, and they began attacking the flames, putting out spot fires, keeping the fire from getting too near the station.

"That's when he pointed out a handful of folks, and we went down and burned out around the guard station proper," continued Vance. "We made sure that held and kept moving on, and it did."

They kept up the battle until after midnight, and with their concerted effort to stop the fire and gradually beat it back, they were able to create a large enough zone so that the station's structures were finally safe. Or as safe as could be expected, given a fire of this magnitude and ferocity, even in the middle of the night.

"There was a sprinkler system going at the time," Tom said, "which was pretty critical. If there hadn't been a sprinkler system, I think we probably would have lost some structures."

By 12:30 a.m., Tom Roach and his core team had been on the fire for more than eighteen hours. The Midewin hotshot crew and the MNICS crew had arrived at the fire later in the day. But Tom, Vance, Pete Lindgren, and many others had been carrying the fight to the flames all day.

Clearly, they were tired.

Greg Peterson, who had also been orchestrating the fight since dawn, was also tired. Right after the station had been secured, Greg drove from the ICP at the Gunflint Lodge and met Tom Roach and others. He was coming to check on them and congratulate them on their efforts. But on his drive up the Gunflint, he had noticed flames not far from the west side of the Trail.

"I know you're tired," Greg said. "And I know your guys are tired. So this is your call. You need to assess whether your team is up for this, or it's too late in the day and people are too frazzled by what's been going on. But we'd like you to do some firing down the Gunflint Trail."

From the station the fire had breached the Trail from the west in a few places down toward Round Lake Road. But there was at least a two-mile stretch from Round Lake up the Gunflint Trail toward the station where the fire was approaching but had not yet crossed the road. If it did cross, Greg knew, it could reach around and catch people working farther up the Trail. So he knew they needed to secure that part of the Gunflint.

By this time Tom and his team were weary, but the adrenalin for the fight was given a renewed surge by Greg's challenge. He and his team did not hesitate.

First, they put together a firing plan. For this operation, he knew he wouldn't need everyone, so he kept back Jeremy Bennett, who became his division supervisor trainee. He kept the Midewin hotshots and three wildland fire engines.

"All other resources we moved back to camp to get some sleep," recalled Tom. "We didn't want to burn everybody out by staying up late."

Before getting to the section of the Gunflint Trail where they could begin doing their burnouts, they needed to literally get through the flames. From the station the Gunflint Trail angles to the south-southeast for about a mile, before turning in a mostly southerly direction for nearly two miles all the way down to Round Lake Road. The first mile of road to the south-southeast of the Trail had been breached. Fire was on both sides of the blacktop, and the wind was still intense.

"With so much wind, flames were hitting the sides of our vehicles when we were driving past that stuff," Tom said. He was in the passenger side of the vehicle, so the direction of the wind was more or less coming directly at him. "My window was rolled up, and I had to hold my hand up in front of my face to shield it from the radiant heat coming through the glass. We had to stop a couple times, because there were flames going across the road as we were trying to get through that stretch. I remember looking in my rearview mirror, and there was a dozer with a transport. It was taking direct flame impingement on the side of the dozer as we were trying to get through there."

The wind was blowing the fire like a bellows, and the heavy licks of flame threatened the vehicles in the caravan.

"We were like . . . holy crap," exclaimed Tom. "It was hot and sketchy getting through that."

Finally, they arrived at the part of the Trail not yet breached, and they began to work.

The wind was still so strong the fighters had to be careful with the burnout. The hotshot squad had flare guns and handheld drip torches. To draw the fire away from the road, they first needed to create an interior fire off the road. Using the flare guns, they set fires well away from the blacktop. Those fires created updrafts that drew the air toward it. With the interior fires burning, the hotshots could use the handheld drip torches to light fire near the road. Once lit, those fires would burn toward the interior rather than away from it.

Tom and his crew worked carefully along the Trail for the nearly two-mile distance all the way to Round Lake Road. It took them two hours to work all the way to the Tuscarora turnoff, but their effort was successful.

"We completed the burnout at three that morning," remembered Roach. "And that held for the rest of the fire."

Finally, after a very long day for almost everyone working on the burnout, they stepped away from the line and bedded down. It was 4:00 a.m.

21 IN THE HEART OF THE HEART OF THE FLAMES

Bob Monehan focused on the water pouring over the south side of his house. He was witness to the heart of a firestorm. The roar was deafening. There was a freight train approaching through the country behind him, to his left and to his right. The whole world was on fire, covered in smoke, ash, and flame.

Strangely, where the water had been pulsing over his property for more than twenty-four hours, the air was relatively clear. He did not cough. He was hosing down the side of his house in the yellow and blood-red light of flame.

When he paused to turn and look across the channel in front of his house, he could see fire across the river through nearly impenetrable smoke. It appeared that everything behind him and to either side was on fire. But here, within this bubble of spray, the flames did not enter. The smoke and ash approached and then moved around on all sides or over, anywhere but into or through the intense bubble of moisture created by the steady thrum-thrum-thrumming of water pulsing over his house and land.

When he looked at the narrows, he could hardly see through the smoke. His boat was moored at his dock, just twenty feet away, gassed and ready. But Bob was not going to leave his place. He had no intention of retreat. Besides, he sensed a retreat up the narrows would be futile. The narrow expanse of water in front of his cabin reached a quarter mile north before closing to a fifteen-foot gap between the peninsula end and the opposite shore. To get into the big open water of Saganaga, he would have to pass through that gap. Bob guessed the entire area was now covered in flame and smoke, or soon would be. He would likely die trying to find it or trying to pass through the gauntlet of a flaming, smoke-choked channel.

The wildfire roar was so intense that if someone were standing beside him, screaming, he would not hear a word. The noise was like some angry demigod, striving to unnerve the foolish mortal that held a puny hose in front of it. The world was one big, grinding whoosh of horizontal wind howl and heat. Bob paused within the path of hungry flames to challenge what was the fire's rightful claim to tinder, dried foliage, roof, and timber.

A normal person—anyone—would feel fear and terror. Rather than stay in front of this juggernaut of flame, most would throw down the hose, make a break for the nearby boat and dock, and take their chances on open water. But Bob was strangely calm and focused on the work in front of him. Many, many years ago he had fought on the front lines in Korea. For one interminable, agonizing year he manned a machine gun nest and bore close witness to the savagery of war. There was a time his machine gun bunker was hit and collapsed all around him. Bob was injured and suffered muscular damage but survived to return home and raise a family. Perhaps it was that experience, seared into his psyche, that now helped him maintain his curious calm.

He had buried his father, who lived until he was ninety-four, and his mother, who passed at eighty-seven. His wife had died seventeen years earlier, and more recently, his partner Charlene. His oldest son had died in a traffic accident a long while ago. Bob was a survivor, and perhaps it was this, too, that provided him with an incongruous sense of grace.

As the fire approached, almost directly behind him now, he turned to behold the flames. Fireballs and ash approached his bubble of moistened air. They approached and like fiery fists knocking on a door, made their roaring presence known. But they did not enter.

The intense wind was blowing in a staggering horizontal blast. The cacophony was deafening.

Bob continued watching. The flames seemed to gather into one mighty phalanx of fire and ash and shoot like a torch into the air. He watched the intense flame rise into the air above his cabin, to the sides. It rose and made a huge arc over his property, where the sprinklers continued their steady thrumming (if he had ears to hear them).

There was a cabin just north of his own, visible from his place, perhaps fifty yards distant. It sat on a higher slab of rock, almost its entire structure visible from the lower saddle of Bob's place. It was dark now. It was a seasonal cabin, and its owners had not yet opened it, and besides, the entire peninsula had been evacuated. No one was nearby. Bob Monehan was alone in the heart of the flames.

Now, watching the arc of fire shoot above and over his property, he was stunned to see it blast down onto the center of the next cabin north. The juggernaut of flame smashed into the shingles and beams and suddenly the entire cabin was on fire. It was as though a huge acetylene torch had been aimed onto the parched and arid timbers, hungry for more food. And there was nothing Bob could do about it, except hold his ground and continue dousing the south side of his house.

Everything was in flames, except this cabin and the near property around it. Bob knew there were other sprinkler systems covering other properties up the peninsula. Michael Valentini and others had been working to get as many of the systems operational as possible. But they were gone now, heeding—like everyone else—the call to evacuate.

Bob wondered if those other properties were experiencing these same phenomena. It would be hours before he could see. He would have to wait for the fire to burn through the country of dry forest and unprotected cabins. It would burn up through the end of the peninsula as though it was a huge fuse wrapped in tinder-dry powder. And then it would jump the Seagull River and continue burning into Canada. The horizontal wind now blasting Bob's backside would easily carry firebrands over the river.

For now, Bob continued watering down the south side of his house. He believed in his sprinkler system. He was confident it would work. But he could not imagine how it would fend off the flames. He could not imagine anything like the torch of flame and ash that rose over his property and fell like a diving menace onto the next cabin over, setting it afire.

The world on fire was truly something to behold.

During the afternoon and early evening before Bob's stand in front of the fire, a steady drone of helicopters and planes made use of the last hours of daylight to course through the blowy sky. They picked up water and used it and retardant to douse the front lines of the approaching flames. Some of the pilots either learned of or saw Bob Monehan making preparations for his postmidnight fight. Some of them, passing over the peninsula in the last light of the day, could see the lone seventy-six-year-old man tending to his pumps and sprinklers. They hoped he made it but knew his fate hung in the balance.

Now, in the early morning of May 7, not long after first light, the fire still blazed to the north. Copters and planes were beginning their grueling day of water pickups and drops. Much of the peninsula smoked from the intense burn that had passed over it in the early morning hours, well before dawn. Fire still smoldered over forest patches nearly destroyed. Cabin ruins still smoked, with red-hot embers still alive and glowing. Occasional patches had escaped the burn. Cabins with sprinkler systems, amazingly, were—like Bob's place—small oases of green in an otherwise burned-over landscape.

At the Valentinis' place and Voyageur Canoe Outfitters, near the start of Sag Lake Trail, were two good examples of the strange juxtaposition of burned-over country and verdure. There was a wooden sign near the road at the start of the Valentinis' drive. A five-foot wooden four-by-four rose out of the ground, with a four-by-four stretched across its top, like the top bracket on an *F*. Hanging from the *F*'s top were two fashionable wooden boards. Across the top board VALENTINI was printed in stylized embossed letters. Below it was a board of equal size that used the same embossed lettering to spell SALLY AND MICHAEL. Because the sign was near the road, it had escaped being burned. But behind it, the brush-covered hillside was ash and blackened to the top of the stony hill.

Later in the morning, when Michael Valentini finally returned, he was disheartened to see the burned-over country at the start of his drive. But when he turned into his five acres, driving down the long drive, he was suddenly and miraculously confronted with the usual green forest of trees and brush.

It was almost the same at Voyageur. Everything outside the arc of spray was burned. Everything within it, green and untouched by the flames.

Most of the peninsula was burned and blackened and looking as desolate as the postapocalyptic landscape of a science fiction movie.

Now, in the early morning light, a lone helicopter droned low overhead. Imagine the pilot's surprise and elation when, in the distance, he beheld a tired and grizzled survivor. He slowed as he approached, dropping still nearer to the ground. Bob stood on his deck, waving to the pilot. The pilot hovered for just a moment, waving down to him, to make sure he was OK. And though Bob was soot stained and weary beyond belief, his smile told the pilot everything he needed to know, and both men laughed.

SUNDAY, MAY 6–MONDAY, MAY 7, 2007, THE BERGS, ISLAND ROAD—As soon as Jim Wiinanen departed the Bergs' place down Island Road the previous evening at the height of the oncoming firestorm, Greg Truex, one of the Bergs' neighbors, walked up the road, carrying a shovel. "Anything I can do to help?" he asked.

Greg had arrived at his cabin the previous week and didn't know until Sunday morning, when the evacuation was called, that there was a fire or that his cabin was in its path. His cabin had a sprinkler system, but—like many others in the area—he had not yet brought it online for the new season. So he stayed. It took him awhile to get it running. After it was operational, he looked down Island Road and decided that for now, at least, he would ignore the evacuation order.

The bay in front of the cabins down Island Road acted as a kind of buffer. The fire was across the bay, moving north-northwest. The Bergs' side of the bay was catching spot fires, as firebrands were flown across the water. But because the area was to the side of the main path of flames, it was easier to manage the occasional spotting, even though the blossoms of flame were numerous, dangerous, and threatening.

The first thing Greg and Ron Berg did was head to his neighbor Lehigh's place. After a short while, they finally managed to get Lehigh's pump running, which is when they discovered the earlier ground fire had burned through some of the piping and destroyed the system. It could no longer carry water to the sprinklers.

In a shed they found 100 feet of fire hose with a nozzle, and eventually, after taking an hour to find the proper connection so they could hook it up to their system, they were able to use it to spray down the neighbor's property, particularly the ground fire that had been doused but was still smoldering.

The area between the Bergs' cabin and their neighbor's was full of dead and downed trees and had no sprinkler system dousing it. While Ron Berg continued spraying the area with water, Keli Berg discovered a fire had bloomed under the neighbor's cabin. Lehigh's cabin was elevated, and below it a firebrand had somehow found its way into a woodpile and in the dry, windy conditions had begun to burn. The fire had not yet grown substantial enough to threaten the bottom floor of the cabin, but it was only a matter of time. "Keli now began carrying buckets of water on a lengthy journey from the lake, up the hill, and down our driveway to the neighbor's driveway to throw water on the flames that were burning *under* his cabin," said Ron.

Demonstrating the kind of bizarre fire behavior that epitomized this blaze, the fire that had crossed Island Road earlier was now back burning down the hill behind Bergs' cabin, threatening to burn a power line right-of-way that was filled with blowdown and thick brush. If it hit that area, the fire would find significant fuel and burn dangerous and wild.

At that moment another neighbor, Mark Lande, walked over the hill with a shovel, and they began talking about what they could do to fight the flames that were approaching the power line right-of-way. Mark's father, Mike, was back at their cabin, keeping a watchful eye on their property and the property of those around them.

Mark and Mike had been helping prepare food at Gunflint Lodge for the increasing numbers of firefighters who needed to be fed. Once they were done making sandwiches, they had somehow gotten around the roadblock and headed up the Trail to their cabin.

"Then Jack McDonnell and his son, Andy, came walking up as well," reported Ron Berg. Jack and Andy had heard about the fire and driven up Saturday morning to get their sprinkler system working. Now there were seven civilians down Island Road fighting flames. They began to talk about pooling resources, wondering if the group could put together enough fire hose to stretch to the flames threatening the debris under the power lines. Jack and Andy soon left and returned with two 100-foot sections of hose. They tried to hook them up, but found one so frayed it was unusable. The other had a hole they repaired by tying a rag around it. Equipped with this 100-foot section and their original 100-foot section, they were able to reach the fire line and douse it.

When Keli and Ron returned to their neighbor's place, they saw a tree on fire, thirty feet up from the shoreline. Every time a gust of wind blew into the torching tree, it sent a shower of sparks into the fuel-rich area between their cabins. Ron trained his hose on the tree and began dousing it with water. It had already been a very long night, but the Bergs knew they could not retire with the burning tree threatening their neighbor's cabin and potentially their own.

"Suddenly a large burning branch broke off the tree," Ron remembered.

"Watch out!" Keli screamed.

Ron managed to step away quickly enough to avoid disaster.

Once they collected themselves, they continued dousing the tree, and after another hour it was finally out.

Ron and Keli headed up the road to check on the others and see if they needed any assistance. Spot fires were still occurring. The wind was still flinging firebrands across the opposite shore to the north and west.

"To our delight, a fire truck was parked just over the hill. While Keli and I had been working on the tree, Sheriff Falk had driven by, and Jack had mentioned that it would be nice if we had a fire truck here to help put out spot fires. Within three minutes GTVFD chief Dan Baumann and volunteer firefighter Rick Johnson showed up with one of the GTVFD fire trucks. They worked until dawn putting out hot spots."

"Rick and I talked about it," remembered Dan. "We followed our training. You'd come up to a structure, triage it, determine if you could save it, and you'd do it. Go to the next one and determine— nope, they got all the wood stacked under the deck, and it's already burning. Let's go on to the next one. You pick and choose what you could do. . . . We just kept knocking it and saving it back. And I'm really happy that every building we worked on was saved."

Finally, the Bergs and the others on Island Road begin turning in, satisfied that for the moment they could rest. It was 4:45 in the morning.

§ 22 SHOCK

Much of the world that had caught fire was still in flames, or at the very least smoldering. To say it had been an arduous night is like saying the D-day invasion was, well, difficult. The D-day invasion was cataclysmic and world rocking. With regard to wildfires in northern Minnesota, the past twenty-four hours had involved once-in-a-lifetime experiences for almost everyone who worked to abate the flames.

The dawn was coming, but it was breaking over a burned-up world. Structures, people's homes, cabins, and businesses, were still standing, but just as many were gone—pillars of ash and smoke in an otherwise blackened world. There were oases of green. These were the places the sprinklers had been running—and many still were. But smoke pushed over the land like an acrid morning blanket of fog.

Group Supervisor Tom Lynch was riding through this landscape in his DNR truck. There were very few people up the peninsula. Almost everyone had evacuated after the sheriff gave the order. And those firefighters who had stayed on and made the last stand at Seagull Canoe Outfitters were all down the Trail. Some of them were getting ready to come back. Some of them had stayed overnight.

Sometime before 4:00 a.m., Sheriff Mark Falk found a place out of harm's way to pull over in his cruiser and try to get some shut-eye. He left the radio on, just in case. He catnapped but not much. By 5:30 a.m. he, too, was on the Gunflint Trail, up near its end.

The only people there were Bob Monehan, who had had a wild ride, and the Seagull Seven, who had also spent the entire night with adrenalin coursing their veins, and GTVFD firefighters Baumann and Johnson. The Seagull Seven had only recently turned in to bed, and the others had already left or were thinking about it.

Through this landscape Sheriff Falk was driving north, nosing up to the end of the Trail to make sure no one was in harm's way. Through this desolation Tom Lynch was driving south, nosing down the Trail to make sure no one was in harm's way.

Both were moving slowly along the Trail, stunned by the acrid landscape, the ashes, so many structures gone, so much devastation. Neither could believe the destruction that had blown through this sacred place. People have been visiting the end of the Gunflint Trail for many, many years. Anyone who has been there knows it is a special place. Its sacred nature was just one of the reasons so many built cabins and homes in the area, and it was home to several businesses—many of them no longer standing.

When they came abreast, they braked. It had been a long night. One of the longest of their lives. They faced each other through the driver's side windows. The sheriff rolled down his window. Tom did the same. They stared at each other for a moment, both shaking their heads.

They could not look at each other long. They looked down. There were no words to utter as the dawn broke on this burned-up world.

When Sheriff Falk looked up, Tom was looking at him.

These were not simple, sentimental men. These men didn't typically weep. But at this moment tears formed in both sets of eyes. They didn't have to speak. There were no words for what had happened up the Trail, for what they had been through, what they had both done, and what they had witnessed. They had a moment of recognition, during which they both knew what the other was thinking. And it was a rare moment, when men come together to do something momentous, something truly exceptional in the face of so much adversity.

After this moment of recognition, they both put their cars in gear and moved slowly forward into the day.

THE PALISADES, SEAGULL LAKE, MORNING—On the morning of their fourth day in the Boundary Waters, the wind finally eased and shifted. The waters, in response, tempered their angry froth. Lee, Layne, and Gus could see the lake's surface had calmed, at least enough that their long-awaited chance had finally come. They broke camp and paddled out.

Around the Palisades point it was still a good two-mile route to their disembarkation point, and the fires were still burning. The primary burn had moved north, but they would be paddling into territory that had already been burned over or was in the process of burning. The air, even east of the fire, was a smoky fog. Ash still fell out of the sky.

Once they disembarked, they would need to hike to Layne's Jeep. They did not expect to see it intact. Or rather, they assumed all their vehicles were scorched hulks. They would have to see.

First, they threaded their way around and through several islands. As they approached the mouth of the Seagull River, they could see the east side of the river burned to a blackened heap, some of the charred wreckage still standing and smoking in the morning's haze. The fire that on the previous night had made its way around the eastern border of the prescribed burn continued to burn north and west, eating through the area where the boat landing, their cars, Trail's End Campground, and more had all resided, before continuing up the peninsula. The end of the Gunflint Trail had been blackened like the tip of a huge spent fuse.

As they paddled up the Seagull River, the east side smoldered. Gus watched a wisp of smoke trail out of a woodpecker's hole in a still standing, blackened tree trunk.

"When we paddled out," Gus recalled, "the only sound was a

white-throated sparrow whistling Oh-Can-ada, Can-ada, like nothing ever happened. Except the bird song echoed ominously across
all the freshly barren land that had been turned into a charscape . . .
a weird echo because there was no foliage to catch the sound any-
more."

On the west side of the river, trees were still catching and firing.
There was a balsam that until now had avoided the flames. But once
it caught fire, the flames spread over the tree quickly, and it burst
into an incendiary column.

The paddlers stared at both sides of the river. They had trouble
reconciling the landscape they were now witnessing with the green
world they had paddled through just three days earlier. They could
not believe it was the same territory. Everything was burned. Except
for the western bank of trees now firing, everything was in charred
ruin. The air was thick with smoke. The day was foggy and eerie and
filled with the sounds and smells of a crackling fire. As they paddled
up the river, it was like they were skirting the edge of a campfire, if
the border of the campfire was the entire forest and everything in it.

They put in at the landing they had left just three days earlier.
After beaching the canoe and kayak, they turned down the road,
hugging the shoreline, moving in the opposite direction from their
cars to take a better look at the carnage. Layne was busy shooting
photographs. They returned to where Lee had earlier pointed out
the five-hundred-year-old cedar that had blown down and subsequently put up new branches from its horizontal trunk, in essence
creating several smaller trees that were anchored by the ancient cedar. Now, after surviving five centuries of burns and blowdowns, it
had finally succumbed to fire it could not escape.

When they returned to their watercraft, a hotshot crew of firefighters was waiting for them.

"Are you Lee?" one of them asked. "We've been wondering
where you guys were for the past few days." Clearly they were interested in Lee, a well-known expert in the area, and perhaps not so
much in his companions.

"Don't I count, too?" Gus asked. All three of their names were on the BWCAW permit.

The others smiled and then told them to get into their trucks, that they would give them a ride to where Layne's vehicle was parked. Over the short route to the parking lot, they saw nothing to dissuade them from the idea their vehicles would be ruined. They were stunned when they turned into the lot and discovered Layne's Jeep, amazingly, intact. Just thirty yards away the forest was a blackened crisp. Layne's vehicle was covered with ash, but definitely drivable, even though it smelled like it had been smoked in a fire pit.

Affixed to the windshield of Layne's Jeep was a note: "Lee Frelich, et al.—please stop in at the Gunflint Lodge." The authorities wanted to speak with him.

Gus smiled at the note. Gus and Layne were well known in their respective professions, as a writer and a photographer. But the real celebrity was Dr. Frelich. "Are you 'et' or 'al.'?" Gus asked Layne. The humor was a fitting end to a trip that had been bizarre, interesting, and unexpected and that would—when the ashes settled—give both Gus and Layne ample material for their own work efforts. For now they were just happy to be heading home.

The hotshot crew escorted them back to Voyageur's lot, where Lee's and Gus's cars were parked. The procession of vehicles crept through the burned-over country at little more than three miles per hour. There were still trees and telephone poles smoldering. They moved carefully along the open road. At any time, some of the weakened, blackened trunks or poles could blow over, either falling on their vehicles or blocking their path.

Finally, they turned up Sag Lake Trail and into Voyageur's lot, once again surprised to find both of their vehicles not only intact but serviceable. In fact, Voyageur had survived the night, like a green oasis in an otherwise barren landscape. Its sprinklers were still running, creating the humidity bubble that—like in other places along the peninsula—had maintained eerie islands of green flora with structures still standing amid the otherwise blackened country.

Once the three were in their vehicles, they drove out, slowly, behind the firefighters' escort. They drove at a snail's pace down the Gunflint Trail, through the burned-over country, much of it still smoldering, the smoke blowing across the land like a dense, otherworldly fog.

At one point they came upon a red fox beside the road. It was lying in the gravel, upright but badly burned. Much of its fur was still intact and red, but its feet were blackened, and it did not move as the vehicles approached. Its muzzle was darkened from the fire. There must be hundreds, thousands of other animals like it who in the previous hours could not flee from the spreading flames and either succumbed or, like this one, were badly burned. It was doubtful the red fox would make it through the rest of the day, as badly burned as it was.

It was a slow ten miles as they drove carefully along the Trail to the Tuscarora turnoff, near the place the fire had started less than forty-eight hours earlier. Most of this section of the Trail had burned the previous night and in the early morning hours, well before dawn. Now it smoldered, with fires still burning. They passed vehicles and firefighters, who waved them through.

Finally, they passed the Tuscarora turnoff, and the Gunflint, while still closed, was at least unburned. They were escorted to the Gunflint Lodge, where there was a kind of media headquarters set up. Here, Lee was the star of the show. The media began to interview him for sound bites to be broadcast later that day.

Gus and Layne both took their opportunity to continue down the road to home. They were actually due home on this day, though they were returning from a trip that turned out quite differently from what they had expected. They stopped in Grand Marais at the Angry Trout Cafe for a well-deserved beer and an excellent late lunch featuring the restaurant's signature whitefish. It was a satisfactory end to their ordeal and a good start to their four-hour trip home to the Twin Cities.

$23 THE FIRST BURNOUT

On Saturday, May 5, Barb Thompson was at her house in Ely, Minnesota, putting in a new flower bed when she was called into the office by her duty officer. There was a fire brewing near Ham Lake, and they might need her help. Since it was drizzling in Ely, she was shocked to hear it. But a short while later she was given her orders, and by morning her bag was packed, and she was flying to Seagull Lake, where Greg Peterson picked her up at Blankenburg Landing.

USFS Firing Boss Barb Thompson began working in fire in 1988. At that time she joined Minnesota's Conservation Corps, during which she helped out with several prescribed burns across the state. After her time with the Conservation Corps, she began working for the USFS as a wilderness ranger, during which she had numerous duties, including assisting with fires in whatever capacities were needed. From 1992 to 1994, she went back to college to finish her degree, while continuing to work seasonally for the USFS. Then in 1998 she became a full-time engine operator for a couple of fire seasons, working out of Cook, Minnesota.

After the 1999 blowdown, she had an opportunity to move to Ely as a fire prevention technician. During the next couple of years, the USFS had a national initiative to bring more people into its fire organization. Many of the fire professionals who had been hired in the 1970s were nearing or taking retirement, and they needed to make sure their firefighting ranks were replenished. In 2001 Barb was able to pick up a permanent position, and she began working in a much more focused way on prescribed burns.

"That's when we started doing a lot of aerial ignition," she recalled. "We worked in the BWCA interior in places that were tough to navigate. For four years in a row we got to work with our Ministry of Ontario colleagues who had been doing large prescribed burns for years. I was able to receive some incredible on-the-job training for aerial ignition on these burns." Most of the burn units (the size of burns on which Barb worked) were between 3,000 and 5,000 acres. One of the burns was 10,000 acres.

When Tim Norman and Greg Peterson called her up to work on the Ham Lake fire, it was because few in the USFS had as much experience and know-how with aerial ignition as Firing Boss Barb Thompson. Since so much of the Ham Lake fire was happening in wilderness that was difficult to navigate, Tim and Greg knew her skill set would be needed. They had also worked with her in the past, so they knew she was familiar with the requirements of working with their team and with the job.

On the morning of May 7, Greg and Barb flew up over the forest around Round Lake. While the heel of the fire had been well tended by Tom Roach and the rest of his Division A, Greg was concerned about the area east and especially south of the forest around the west end of Round Lake. If the wind shifted, the smoldering fire could burn south in the BWCAW, which could potentially create a catastrophic southern BWCAW wildfire. Similarly, if it burned back east, Tuscarora and other structures down the south side of the Gunflint Trail could be threatened.

Once up over the forest, Greg showed Barb the area he thought should be burned out. If Barb burned it correctly, it could create a blackened buffer zone of spent fuel that would prevent any fire from moving south or east.

"We flew over the area looking for any pinch points, where we could connect to a wetland or lake. We were looking for patterns in the terrain we could use, like a portage or something similar," Barb remembered. Barb was also examining fuel types. This area of the forest was mostly heavy, mature timber. There was black spruce, some blowdown, aspen, and balsam fir. To Barb, who had been working in the area for several years, it appeared to be comprised of typical boreal forest components. This kind of forest burned more slowly than, for example, dry grasses and jack pine. And she noticed it hadn't been burned, so it was much denser and closed in.

Aerial ignition is an art that involves numerous players. First and foremost is the aerial ignition firing boss, who based on a wide variety of factors—fuel, terrain, weather, and more—develops an aerial ignition plan. Once the plan is reviewed and approved, the aerial ignition firing boss executes the burn with his or her team. Aerial ignition is executed from a helicopter and requires a firing boss, a pilot, and a plastic sphere dispenser (PSD) operator. The firing boss sits near the pilot. The operator is in back, managing the PSD. And the entire team has microphones and is able to communicate. Communication, given the complexity of dropping what in essence become fireballs from the sky, is of optimal importance.

The plastic spheres are ping-pong-sized balls filled with high-grade potassium permanganate. With the helicopter door open, the PSD operator can drop the balls to the ground. Before the balls (sometimes referred to as "ping-pong balls" or "spheres") are discharged, the PSD injects them with ethylene glycol (antifreeze). Approximately twenty to thirty seconds after being injected, a chemical reaction causes the ping-pong balls to ignite. The PSD operator controls the mechanism from the back seat, while the burn boss oversees the operation from the front.

"We were about 200 to 300 feet above ground level," recalled Barb. "I picked out a firing pattern that I wanted according to wind direction, fuel on the ground, hazards . . . and you have to do it carefully so you don't end up smoking yourself in."

Midmorning on May 7 the wind was still strong, so managing

the aerial burn became a slow, deliberate business. Barb and her crew would work a line from south and west of Round Lake over to Brant and Gutter Lakes. There were natural marsh breaks south of Brant and Gutter, where the fire would presumably stop. Throughout the morning and midday they would drop a strip along that line and then wait for the timber to catch fire. Once they had a clear idea of the fire's behavior, they could make any necessary adjustments to their burn pattern and continue dropping.

Because the ping-pong balls needed to be very carefully placed, there were a series of commands Barb used with her PSD operator. When they reached their desired elevation and east-to-west line, Barb radioed her operator.

"Start your machine," Barb said.

Once it was started, the operator confirmed, "PSD operational."

The helicopter pilot was not stationary but moving in a direction that facilitated the fire line below. The ping-pong balls were fed into the PSD continuously, but the drops were executed by the operator so that they hit the forest below, the fuel source, at exactly the right location. Since the copter was moving and the wind was up, the drop had to be carefully timed—like leading a bird in flight with a shotgun.

When the helicopter reached the correct location, elevation, and speed, Barb ordered the drop. "Start firing."

Then well before the length of the desired firing strip was reached, Barb had to make sure to alert her operator to the impending cutoff. "Prepare to stop."

Once the exact location was reached, she gave the order to "Stop firing."

Using the ping-pong balls, Barb's team began laying down east–west lines in the dense forest around and west of Round Lake. Because they were setting fire to the forest in an area difficult to navigate on the ground, they needed air support to ensure that this fire didn't burn out of its boundaries. Consequently Barb was constantly in touch with the corresponding air attack.

From the southwest corner of Round Lake there is a 77-rod portage to West Round Lake. "At one point I remember the fire line slopped over the portage," recalled Barb. "I know the air attack had a hard time dealing with it." But eventually they were able to paint a line that stopped the fire's progression south.

On this day Barb and her team successfully burned 900 acres of forest, creating a blackened barrier that held the southern line through the rest of the Ham Lake fire. They were finished by around 4:00 p.m. It was a good test for what was about to happen on Tuesday, day four, of the fire, as the Ham Lake fire continued its pinwheel burn across northern Minnesota and southern Ontario. Barb did not yet know it, but the next day she and her team and more than four hundred firefighters led by Greg Peterson, Tom Roach, and others would be tasked with stopping a potentially devastating turn in the fire from destroying numerous cabins, resorts, and structures around Gunflint Lake.

Sometime during the morning of May 7 Forest Supervisor Jim Sanders finally made the call. Conditions throughout Sunday night and early Monday morning had worsened, and the fire had grown and would soon be crossing the international border. It was only a matter of time. In spite of the fact that elevating the approach to the fire from a Type II incident to a Type I was never done easily, by this time Sanders was able to make the call with confidence.

The MIFC information staff issued the first of two May 7 press releases about the fire. "The Ham Lake fire continued to burn actively through the evening Sunday and presently is 16,500+ acres," the release stated. "It has burned into the area at the end of the Gunflint Trail where severe damage to multiple structures has occurred. The National Weather Service is forecasting severe fire conditions for the area this afternoon and tomorrow. South winds will switch to the west."

The release also stated that "a Type I Team from the Great Basin has been ordered and will begin arriving today. They are individuals from Utah, Nevada and southern Idaho."

The morning release stated that there were "150+" people working on the fire. By the afternoon's press release, that number had grown to "200+" people.

The Type II team had been called up from MIFC. The Type I team was called up from NIFC, the National Interagency Fire Center in Boise, Idaho. The Type II team call-up was a state response, while the Type I call-up was a national response. Many of the firefighters who worked on the Ham Lake fire in the first three to four days also worked on national fires, so they understood that the difference between a state and a federal response was more than a matter of degrees. And because most of the personnel and resources from the Type I team would be coming from the West, it would take at least the next two days for the team to get into place. During those transition days, Greg Stegmeir and the rest of his Type II team would continue fighting the fire.

In essence, the elevation of an event from a Type II to a Type I approach involved more than just a doubling of resources. According to the Federal Emergency Management Agency, a Type I event is the "most complex, requiring national resources for safe and effective management and operation." Typical of this category of event, "operations personnel [will] often exceed 500 per operational period and total personnel will usually exceed 1,000." The incident commander would be Paul Broyles, national fire operations chief for the National Park Service and the IC for the Great Basin National Incident Management Team. He would bring with him an extensive team and resources. Perhaps most important, resource requests for a Type I event were first in line when addressing disasters. In practice, that meant if, for example, another fire broke out in some other place in the nation, and a Type II team in, say, Mississippi requested resources from NIFC, there would be a queue for those resources, and the Type I team would be first in line.

THE BIG BURNOUT

§24 BACKFIRE

The first few days of the Ham Lake fire confirmed the key, pivotal role that weather played in the blaze. Higher than normal temperatures, low relative humidity, near-drought conditions in areas containing plentiful fuel, and high winds made what could have been a manageable wildfire into what was becoming one of the worst fires in Minnesota history. Unfortunately the fire weather would only worsen, which would have a profound impact on the days ahead and on the choices made to combat the fire.

On Sunday, May 6, meteorologist Byron Paulson was building a sauna at his cabin outside of Cloquet, Minnesota, when his sister-in-law stopped by and told him about a wildfire that had been reported during the morning news. Byron knew that for a fire to be reported, it must be significant. And because Bryon was an IMET (incident meteorologist), he thought he should check on it.

For many years, IMETs were key members of the western firefighting teams, but it wasn't until 1987 that they decided to nationalize the program. "When I started in 1987," noted Byron, "there were maybe thirty of us. There were only four IMETs in the eastern United States." Today there are around eighty nationwide.

At the time, Byron was a lead weather forecaster at the National Weather Service (NWS) in Chanhassen, Minnesota. But having grown up in northern Minnesota and having spent plenty of time in the woods, he had a natural affinity for wilderness areas, which is in part why he also became an IMET.

To become an IMET, NWS meteorologists must go through specialized training and become certified. Once certified, IMETs support federal, state, and local firefighters working on the front lines. Of course not every fire becomes significant enough to require an IMET. But once a fire requires a Type II incident response, an IMET is usually on site and becomes a key member of the incident command team, issuing near- and far-term weather forecasts that command teams use to help keep fire crews safe.

Unfortunately, cell reception at Byron's cabin was spotty. Sometimes he could get connected on a call only by climbing a ladder to its top rungs. On Sunday, after hearing the news about the fire, he climbed a ladder and called the MIFC. Sure enough, they told him the Type II team had requested his services. It was already late in the day on Sunday, so he began making the necessary arrangements with the home office in Chanhassen, letting them know he would be there first thing in the morning but would be heading north later that day.

Once in the office on Monday morning, one of the first things he did was analyze the weather for northern Minnesota, particularly the region being affected by the fire. With his near-term forecast in tow, he gathered together the rest of his weather forecasting equipment (he needed to maintain a small office at the fire) and headed north. In the late afternoon he checked in at the ICP at the Gunflint Lodge and issued his first weather forecast. Not only was the news bad, but the long-term forecast—which he had not yet documented—appeared equally ominous.

On Tuesday morning he compiled the day's forecast in time for the 7:00 a.m. briefing. And while he did this on May 8, he was already keeping tabs on a cold front that he expected to move into the region sometime around May 10. He was also well aware of the wind shifts, which had pushed the fire largely west the first day but then gradually began turning in a clockwise fashion approximately twenty degrees or more per day.

Later that morning he compiled a weather risk assessment for Tim Norman. In addition to general seasonal considerations and more specific humidity, wind, and temperature readings for the first few days of the fire, Byron noted:

> The approach of a cold front as the high pressure moves into the eastern Great Lakes is a critical weather situation to watch. This will bring a period of gusty southwest winds of around 20 mph with higher gusts. Relative humidity is often still in the 20% to 30% range in the transition to southwest winds. Typically, a day of gusty southwest winds is followed by the passage of the cold front bringing an abrupt shift to gusty northwest winds. . . .

The approaching cold front, Byron told Tim and others at ICP, was definitely worrisome and could have a significant impact on the fire.

Much of Tuesday was spent moving the ICP from the Gunflint Lodge back to the Seagull Guard Station, which was now secure. But also during Tuesday Byron was having conversations with the incident command team, particularly Greg Peterson, about the coming change in the weather. Byron let Greg know that for much of May 8 the winds would continue coming out of the west-northwest, which if unchecked would push the fire into the Magnetic and Gunflint Lakes area. Peterson and others knew that threat was serious, because there were resorts and cabins on the west side of Gunflint and Magnetic Lakes, and the south side of Gunflint Lake. Some estimates put the value of those structures at approximately $22 million.

Once Greg considered the oncoming fire and its approach from the west-northwest, he knew they had to do something about it.

A sure sign of the growing importance of the Ham Lake fire was the appearance of Minnesota's governor. On the morning of May 8 Governor Tim Pawlenty did a flyover of the area to review the devastation and loss that had already occurred at the end of the Gunflint Trail. Sometime after eleven o'clock, he landed on Gunflint Lake, disembarked at the Gunflint Lodge dock, and was taken up to the open area in front of the nearby GTVFD Gunflint Fire Station 2. From that location, the smoke plume was widening in the distance behind him, which formed a memorable backdrop to his press conference.

His remarks were widely reported in the press. On Monday he had put the National Guard on standby. And on this morning he said, "We've told them anything they need from us—helicopters, people, personnel, equipment, money—we'll be happy to make anything available that they need."

On the evening of Tuesday, May 8, Tom Roach received a radio call from Greg Peterson. "Hey, Tom, I'd like to meet up with you," Greg said. "I have a proposal for an assignment that I'd like to talk to you about for tomorrow's operational shift."

Over the next hour Tom and a handful of other key participants met with Greg, who had a plan to combat what everyone thought might be an approaching firestorm.

At this time the fire was still a couple miles to the west-northwest of the Magnetic and Gunflint Lakes area. Since the wind was continuing to turn in a gradual clockwise formation, it would likely push the fire into the area, threatening everything to the west and south of Gunflint and Magnetic Lakes.

In an odd kind of fortuitous way, ahead of the approaching cold front there would be a ninety degree shift in wind, with the breeze anticipated to come out of the southwest. Unfortunately, Byron Paulson had told them, as the cold front passed through, the wind would again shift and strengthen, coming out of the north-northwest with so much force that a red flag warning would likely need to be issued. Byron estimated wind gusts on Thursday, May 10, of up to 30 miles per hour.

Greg asked Tom and his team to scout the area and see if they could put together a plan to secure fire lines using existing roads, trails, and whatever else they could along a line that ran west of Magnetic and Gunflint Lakes to the southwest all the way to the

Round Lake Road turnoff on the Gunflint Trail. If they could create a secure enough line, they could use ground and aerial firing to burn a huge section of forest in front of the oncoming flames. Ideally, they would wait until the wind shifted out of the southwest, before the approaching cold front, so they would have the advantage of the wind pushing the fire in a north-northwestern direction. If they could burn a large enough section of ground in front of the coming fire, they could prevent it from burning the west ends of Magnetic and Gunflint Lakes and continuing along the southern Gunflint shore, on which the historic Gunflint Lodge resided, as well as other resorts and many other cabins.

Greg had a map that showed the two-to-three-mile Forest Service road 1347. That road ended at the long peninsula marking the narrows running from Gunflint Lake into Magnetic Lake. From the narrows the road ran almost three miles due southwest to the Gunflint Trail. The map also showed a large gravel pit approximately a quarter mile from the Gunflint Trail end of FS 1347. Depending on the width and quality of the road, the gravel pit area, and whatever else they could find west of the gravel pit, there was a chance they could put in a wide enough fire line in front of the southwest wind shift. Once the wind shifted, they needed to be in position to begin the aerial and ground firing.

"If the main fire burned into that area," recalled Tom, "there really wasn't anything to catch it after that. We really had that one option, to essentially make our own fire in front of the main fire. Peterson wanted us to see if we could tie the fire in from the west end of Magnetic Lake and burn it down a series of hand lines or dozer lines or whatever else we could find or build until we got down to the Gunflint Trail. He wanted us to put in a big line of fire that would act as a kind of catcher's mitt for that main body of fire burning into it. The goal was to try and get them to suck together and then, choked of fuel, stop."

By dusk on Tuesday, May 8, Tom and his partners were cruising along FS 1347, checking out the area due west of the south end of

Magnetic Lake all the way southwest to the gravel pit and Gunflint Trail. During their reconnaissance they discovered an abandoned power line running from the gravel pit almost due west for approximately a mile to where Round Lake Road turned off the Trail. The power line poles were gone, but the ground beneath where they had once stood was largely cleared. Tom, Vance Hazelton, Jeremy Bennett, and others thought that—with an enormous amount of work—they could make it into a line that would hold.

There was an area due west of Magnetic Lake that had no road and was too steep and rocky to use a dozer to build a line. For that section a crew would need to build the line by hand. It was rugged country, but once completed the hand line would tie in from the southwest shoreline of Magnetic Lake to FS 1347. For the rest of the line down to the gravel pit, they could use dozers and crews to widen the road. Then from the gravel pit due west along the abandoned power line they would again use crews to clear debris and widen it into a swath of ground cleared of fuel.

As the sun settled into the western horizon, everyone who looked at what they were planning to do knew it would be an enormous task. They would need at least two hundred firefighters and several pieces of specialized equipment they didn't yet have on site. And they would be working through extremely rugged country. Parts of the wilderness west of the Gunflint and Magnetic Lakes were heavily boulder strewn and marked by granite ridges and lowland swamps. Putting lines through the area and conducting a backfire was going to be tough, to say the least. But after they reviewed the plan with Greg Peterson, he agreed they had no choice but to forge ahead and try.

Late on the evening of May 8 Tom began requesting resources. "We got back that evening and told Greg we can do it and here's what we need," recalled Tom. "Four to five crews, a couple strike teams of fire engines, two to three strike teams of track equipment, two to three strike teams of dozers . . ." A strike team of wildland fire engines is comprised of five engines, so Tom was requesting ten to fifteen engines. A strike team of tracked equipment and dozers is comprised of two vehicles, so Tom was requesting four to six tracked vehicles and four to six dozers.

Tracked vehicles are hybrids that run on tracks, have large water tanks with water pumps and hose, and can go where fire engines and dozers cannot. They are a kind of hybrid engine/dozer. A crew is comprised of approximately twenty firefighters, so in addition to those he already had on hand, he was requesting approximately another hundred fighters.

By the end of the evening Tom estimated he had requested approximately two-thirds of the total resources he would need to get the job done. He felt certain that by the time they all got on the ground and began working the plan, he would be calling in additional resources. But for now he would have to wait to see what else would be required.

Greg began working with the plans leader and others to fulfill Tom's requests.

On Wednesday morning, May 9, everyone was up early. Tom Roach was already speaking with his team members and others to determine next steps. Byron Paulson was compiling a weather forecast for the 7:00 a.m. briefing. Barb Thompson was already contemplating an aerial firing plan. Everyone knew it was going to be a momentous day. They were either going to succeed and put in a three-mile firing line, in front of which they could burn off the ground and backfire a patch of fuels so large it would stop the main fire in its tracks. Or they were going to fail, and conceivably $22 million worth of resorts and cabins might burn.

Because everyone knew that much rested on the day's events, they grabbed whatever breakfast they could, knowing they might not get another chance to eat once the full throttle of the day was under way.

At the 7:00 a.m. briefing the Type II leadership team covered the previous day's efforts and the plans for the day. Greg Peterson outlined the huge backfire they were going to attempt and began identifying who was responsible for specific parts of the effort. While the big backfire was the most significant activity for the day, consuming the largest percentage of equipment and personnel, it wasn't the only part of the fire that needed to be managed. During the briefing, John Stegmeir, Peterson, and others outlined everything that would be happening for the day.

Since this was a transition day, with some of the Type I teams beginning to trickle in, Type I team Firing Boss Mike Fralicks was introduced and briefed on the effort. Mike would be leading the entire burn effort. Barb Thompson would be managing the aerial burn effort for Mike. But since Mike was from the West, he had just arrived and was unfamiliar with the area and terrain. And since Barb had done many burns in the area, the last one just two days ago, she would be a key resource in getting the burn done.

One of the most important parts of the briefing was communication. Everyone had radios and was alerted to the proper frequencies and channels. Throughout the day, when so many were spread over such a wide area, with some overhead in numerous types of aircraft, it was going to be important to maintain clear lines of communication.

At the meeting all that had been covered, including Byron Paulson's weather forecast, was rolled up into the day's IAP and distributed to those who needed a copy.

Sometime later in the day, Byron had noted, the winds were going to shift and begin blowing out of the southwest. When they did, Greg, Tom, Mike, Barb, and many other firefighters wanted to be ready with their fire lines and aerial and ground firing. Once the wind shifted, the real show would begin.

At 7:30 a.m. the main briefing ended, and each division broke off to have separate meetings. At this time Tom moved his team leaders to one side and began reviewing the day's operation. All of Tom's Division A lieutenants were present, including tracked equipment bosses, dozer bosses, and crew bosses. Once everyone was assembled, Tom detailed the work ahead for each of his bosses. Along the line from the southwest shoreline of Magnetic Lake all the way down to the gravel pit and then west along the abandoned power line, firefighters, equipment, and supplies would be spread out in a massive effort to build a solid firing line.

At 8:30 a.m. Tom, Mike, Barb, and a handful of others met with Greg to have one final review of the day's key, crucial operation. None of the main players knew it, but at this point there was so much work to be done over such a wide area it would be the last time they all met in person. The absence of at least one more face-to-face meeting and an unexpected breakdown in communications would later result in a mishap that threatened for a time the entire operation.

Tom spent the day with his crews and engines, working up and down the myriad parts of the line. They needed to limb and thin as much vegetation along the road and other lines as they could. There were approximately two hundred firefighters on the ground with chain saws, axes, shovels, and other tools making and clearing lines. Whenever possible, they were supported with dozers and tracked vehicles. In several places along FS 1347, the dozers needed to carve out parking areas, places along the road where, for example, a 3,000-gallon water tender could pull off the road if it ran into an oncoming fire engine or another water tender. Throughout the day they also needed to coordinate traffic up and down the road, making sure people coming in and going out would not meet and get stuck in a bottlenecked road.

Tom was in constant radio contact with his crew and equipment bosses, making sure the hand lines and road lines were getting cleared, widened, and ready and that the crews had everything they needed to get the job done. Massive amounts of hose were being laid, with pumps running down into nearby water sources.

"We had several layers of contingencies for LCES," explained Tom. LCES is a firefighter acronym that stands for lookouts, communications, escape routes, and safety zones. "We're going into a narrow road with a lot of apparatus. If fire slopped over and we had to get everybody out, we didn't want people to get caught in there."

As part of this effort, Tom assigned an entire team to set up sprinklers on all the lakeside cabins on the west end of Gunflint and Magnetic Lakes. There was also a renowned group of cabins on Gallagher Island on Magnetic Lake. Some people referred to the island as Million Dollar Island, or the Gnome Island: the Million Dollar Island because of the luxury northwoods cabin and other structures on it, and Gnome Island because the owner had posted gnome statues all over the island, facing out—similar to the stone statues on Easter Island.

One other LCES effort involved Dan Grindy. Since the fire had burned through Dan's Division B territory two days ago, he had been reassigned to Tom's effort. Dan and his team commandeered a fleet of sixteen-foot aluminum fishing boats. The boats were outfitted with pumps and hose and were positioned up along Million Dollar Island and the western shore of Magnetic and Gunflint Lakes. If the unimaginable happened and firefighters needed to escape the oncoming flames, they could retreat to the safety zones of the sprinklers covering the lakeshore cabins. From there they could be evacuated using Dan's boats.

Barb Thompson went back to the ICP and wrote up an aerial ignition plan. Since two helicopters would be doing the aerial ignition, she had to coordinate the effort between the copters and the helibase, which would be refueling, supplying the PSD operators with firing fuel (ping-pongs), and supporting them. Once the draft plan was finalized, Barb and Mike Fralicks needed to review it with Forest Supervisor Jim Sanders. "If we lost this," Tom explained, "we just introduced fire a mile and a half from where it was, right on the brink of structures." Supervisor Sanders would need to sign off on the plan, since it carried inherent and serious risk.

Before lunch on May 9, Tim Norman, John Wytanis, Jim Sanders, and others were in Grand Marais at the USFS office, briefing Paul Broyles, the Type I incident commander, and several members of his team. They discussed the firefight to date and what the fire was expected to do in the near future. After this briefing everyone headed to the Gunflint Lodge. From the docks and deck at the Gunflint Lodge, there was a clear view of the oncoming fire and smoke plume to the northwest, and the work being performed by the numerous crews.

As the day progressed, Greg, Tom, Barb, Mike, and others began waiting for the anticipated shift in wind direction. Tom and his two hundred-plus person team with their accompanying equipment, sweat, and hard labor were gradually getting the necessary fire line built, widened, and set up for the show they knew would happen later in the day. It was a massive effort, but as the day continued, it appeared more and more likely they would be ready on time.

At around ten thirty that morning Byron Paulson sent up his first pilot balloon, known as a pibal. The pibal is a small balloon filled with 10 grams of helium. Weather forecasters use pibals and a theodolite instrument to determine wind speed and direction. The theodolite is used to measure the balloon's angle on a horizontal and vertical plane. The pibal is released, and knowing its speed of ascent and using the theodolite's measurements, the forecaster can get an accurate sense of wind speed and direction along the vertical column through which it rises.

"Sometime during the night, a weak little weather disturbance dropped out of Ontario and turned the winds around so they were coming out of the northwest," recalled Byron. "I was confident they were going to switch back to southwest during the day, but it was pretty dicey in determining how quickly they were going to change."

After that first balloon at midmorning, he began launching them every hour. After he calculated the readings, he would contact Greg and update him on the wind shifts. "You could see by noon and a little later it was starting to turn," noted Byron. "It was a day that had relative humidity that was pretty stinking low. Even in the evening hours. All the way until 10 p.m. the humidity stayed below

30 percent. So they had real favorable conditions in terms of wind direction, speed, and humidity to get that burnout wrapped up on the ninth, ahead of the strong trough coming in the following day."

At around four o'clock Tom believed they were ready. Peterson had been overhead in a Beaver plane most of the morning and afternoon, checking on the backfire as well as several other areas where firefighting was under way. Greg radioed Tom.

"Are you guys ready to ignite?" he asked.

"I think we are," Tom said.

"OK. Hold off. The winds aren't yet right."

During this time there were two heavy helicopters and two heavy air tankers in a constant rotation, dipping into Gunflint Lake, filling their tanks with water, and then making drops along the south side of the firing line. "They were just pounding the green, keeping it wet," remembered Tom. "We were trying to make sure anything that spotted didn't have receptive fuel there. So it was an air show overhead, with air attack directing it. Plus the Beavers are up. There were probably seven to eight aircraft over that incident the entire time."

Finally, sometime shortly after five o'clock, Byron let Greg know the winds had shifted out of the southwest. At 5:15 p.m. Peterson radioed Tom, Barb, and Mike and let them all know they should start firing.

"We got confirmation at 5:15 and began firing at 5:20," Tom said.

Minnesota's oldest living tree, an approximately one-thousand-year-old cedar on Three Mile Island, Seagull Lake. Photograph copyright Layne Kennedy.

Professor Lee Frelich, overlooking the smoke-filled air across Seagull.

Photograph copyright Layne Kennedy.

A massive flame explodes near Gunflint Lake.

Photograph by Richard Sennott. Copyright 2007 *Star Tribune*.

Aerial perspective of Ham Lake fire.

Photograph by Jody Leidholm, MNDNR Air Tactical Group Supervisor.

The fire burns close to a cabin protected by sprinklers.

Photograph by Lee Johnson, USDA–Superior National Forest.

Fire taking out power lines along the Gunflint Trail.

Photograph by Minnesota Incident Command. Courtesy of Tom Roach.

Stephen George Posniak, whose fire started the Ham Lake fire.

A helicopter drops water on the Ham Lake fire. Photograph by Richard Sennott. Copyright 2007 *Star Tribune*.

Firefighters create a backburn along a section of the Gunflint Trail. A drip torch is on the ground to the right of them. Photograph by Chief Dan Baumann, Gunflint Trail Volunteer Fire Department.

Structure blaze. Photograph by Chief Dan Baumann, Gunflint Trail Volunteer Fire Department.

Hillside ablaze during the Ham Lake fire. Photograph by Marlin Levison. Copyright 2007 *Star Tribune.*

Half of a Kevlar canoe at the end of a row of canoes. Photograph copyright Layne Kennedy

Burned fox.

Photograph copyright Layne Kennedy.

One of the Minnesota DNR's CL-215 water scoopers drops water on the fire.

Photograph by Minnesota Incident Command. Courtesy of Tom Roach.

Firefighters in a sixteen-foot aluminum fishing boat, commandeered for battling the fire from the water. Photograph by Stefanie Mojonnier.

A firefighter contemplates the ashes, refuse, and stone remaining from a burned home.

Photograph by Brian Peterson. Copyright 2007 *Star Tribune*.

Two weeks after the Ham Lake fire. Photograph copyright Travis Novitsky, www.travisnovitsky.com.

§ 25 FIRE DOWN THE LINE

By 5:15 p.m. Tom Roach, Mike Fralicks, Jeremy Bennett, and Barb Thompson had all received word they could begin firing. At this point Tom's role was division supervisor for everything being done on the backfire. Essentially, he was responsible for the entire effort. Jeremy was his trainee in charge of holding the fire line. In that role Jeremy had full responsibility for holding the line from the southwest shoreline of Magnetic Lake, all the way southwest to the gravel pit, and from there due west to where Round Lake Road turned off the Gunflint Trail. Mike was responsible for firing along the line, and Barb was responsible for aerial firing.

The plan discussed after the morning briefing, which was the last time Tom, Jeremy, Mike, and Barb had a face-to-face meeting, was that once the lines were finished and the winds became favorable, Mike would oversee the hotshot crews on the ground as they began ground firing along the southwest shore of Magnetic Lake, working southwest along the hand and dozer lines until they got to the FS 1347 line. Then they would continue burning along FS 1347 to the gravel pit. Coincident with the ground firing, Barb would be leading the aerial firing farther into the interior, away from the line. Barb's firing would create the suction needed to ensure the ground-fired line would be pulled toward it, creating the burned-out area that would stop the oncoming fire front from continuing south.

Barb's prep work with the helicopter prevented her and her crew from getting into the air until 6:00 p.m. When she was airborne, she radioed Firing Boss Fralicks to check in. Mike was along the west side of Magnetic, working with the hotshot crews on the ground-fire ignitions.

"His radio began making noises I had never heard before," recalled Tom. "It just fritzed out."

At that point Mike's radio became inoperable, and he lost contact with Barb, preventing her from communicating with him on her aerial burn. Since Mike was in rugged country, it would take awhile to get him a new radio. Unfortunately, Barb did not have the luxury of time, so she began dropping spheres due west of Magnetic Lake, following her aerial ignition plan.

"That burnout was a little tricky," recalled Barb. "Although most of the [burn area] was pretty straightforward and consisted primarily of grass with a component of jack pine—as we had burned it a few years prior—there were a lot of smoke concerns, from hand ignition that was also occurring at the same time on the ground. And we were very concerned about spotting across the Canadian border, Million Dollar Island, and Gunflint Lodge."

At this time Tom was along FS 1347 in a lookout position. "We had fire on the ground already, and we were looking for heat in the interior to start pulling our flames toward it," remembered Tom. Initially, that was exactly what happened. The interior fire Barb created from her aerial firing effort drew the ground fire toward it. The plan was working. The firing crews worked down the lines, firing as they went, and then holding the line and making sure none of their fire or Barb's interior fire slopped over their line and started burning forest behind them. Once the line was held, they would pass along the firing tools to the next hotshot crew farther down the line, and they would begin setting ground fires and holding the line.

For the first third of the line, approximately a mile, the plan worked perfectly.

"It was a very slow pace to light off the ground because everything immediately was right up in the trees," recalled Tom. "So you had to go very slowly. And if it had slopped across . . . the whole operation could have lost the egress back out of there."

At approximately 6:10 p.m., Tom watched Barb and her helicopter crew head farther to the southwest. He wondered about it but then thought, *she's doing a loop into the interior, and she's going in to light more suck heat. She's putting some hotter fire inside that'll draw this fire from the edge.*

But Barb, whose vision of the ground was probably impaired due to the heavy smoke in the area, ended up farther to the southwest, dropping fire in front of the west end of the firing line, near the Round Lake turnoff.

Earlier, Jeremy, as the holding boss, had traveled down near the west end of the line. His crews were still widening and clearing ground along the old power line, and he wanted to make sure that by the time the ground fire reached them, they would have the firing line in the best possible shape.

At around 6:15 p.m. Tom received a radio call from Jeremy.

"Hey, Roach, we need to talk right now."

"What's up?"

"The helicopter is way down past us dropping fire, and there's no way we're going to have people down there to hold that."

"What?"

"Yeah. And it's only a couple hundred yards off the line. The aerial fire's going to get here before your guys do with their ground ignition."

Suddenly the burnout had become dangerous. If the fire burned down to the line and there was no one there to hold it, or there was no ground fire set in front of it to consume fuel in front of the oncoming flames, it could skip over the line and begin burning to the south and east. And if it burned far enough south and east, firefighters could be trapped in the center of a firestorm.

By 6:30 p.m. Tom had considered his options and come up with a new plan.

Barb had run out of firing fuel and returned to the helibase to reload.

Tom radioed Greg.

"I want to shut down the aerial ignition until we're ready for it," he said.

Once Tom explained, Greg agreed, and for the moment, Barb remained on the ground.

"The aerial burn was one and a half miles ahead of us," remembered Tom. "We weren't anywhere near where we needed to be to handle it. We were somewhere down around a mile into the burn when Bennett's call came in. We were going to burn to the gravel pit, and then the plan had been we'd keep going down the power line."

But now, Tom recognized, "If the fire burned over the road and there was fire on both sides, we can't get 150 people out of there. All of a sudden it was a nightmare situation."

Tom radioed Jeremy and told him to get all his resources down to the end of the line and start ground firing ahead of the approaching flames. He told Jeremy to fire and hold the line from the Gunflint Trail and then move east toward the gravel pit. Tom would make sure Mike's crew continued firing to the southwest, and they would meet Jeremy at the gravel pit.

"You never burn from two directions and meet in the middle," Tom said. "You just don't do that, but we had to. We had to make a quick decision and flip-flop the operation and burn it toward the middle from two different directions."

For the next hour Jeremy worked the west end of the line, and the fire they laid down was, thankfully, pulling in nicely. Gradually, they worked their way to the gravel pit. Once they were certain they had it under control, Barb returned to her aerial firing work, laying down strips of interior fire in just the right places to suck in the ground-fire flames.

"There were a couple spot fires that jumped over on the Magnetic side of the line," remembered Tom. "From six o'clock to eight o'clock there were two times we had spots we had to secure using helicopters and hotshot crews. They were in heavy blowdown but, luckily, just a stone's throw off the road."

By 9:30 p.m. everything was going so well they released some of the crews working the line. And at 10:00 p.m. they called the ignition complete and began loading up vehicles and equipment and releasing more firefighters so they could go get some badly needed rest.

"As we finished up the burn around ten," recalled Greg, "that's when we turned it over to the Type I team."

"The Type 1 team as well as many dignitaries from the forest and state had arrived and were at Gunflint Lodge," remembered Barb. From the lodge, as the afternoon and evening came on, they were able to watch the entire effort across the lake to the north and west unfold. "Thankfully, I didn't know I had an audience until after the burn was completed."

The Type I team and everyone else looking from the lodge that late afternoon and into the evening must have been impressed by what they saw.

In retrospect, Tom Roach remembered the daylong effort as one of the most intense of his entire firefighting career. It wasn't only the Herculean task of building a three-mile-plus secure-worthy fire line in the span of ten hours, but it was also about ramping up their resources, coordinating the myriad aspects of the effort, and doing it all in a climate when there were plenty of distractions. "I kept getting calls about power lines being threatened down the Gunflint Trail to the north of us. And that was a big deal, because earlier they had burned out and the whole ICP had to move."

The importance of the power and communication lines was compounded because of the time of year. No one up the Trail ever appreciated losing power and their phones. But in two days lots of people were going to be flowing into the area for the fishing opener. For resorts up the Trail, this was a huge and important weekend. People with reservations were wondering if they could still come and fish, and if they could not, the resort and outfitter revenue lost would have a serious impact on these businesses.

Around 4:20 p.m.—at the height of their effort and during the middle of the burnout briefing, with the start of the burnout only one hour away—Tom began receiving calls about the power lines being threatened. "But we couldn't do anything about it," recalled Tom. "By 4:30 in the afternoon we were getting calls that if we could have dedicated the needed resources to, would have compromised our whole burnout operation." It was a distraction that prevented him from touching base with Barb one last time. "There were just so many things happening. The power lines were just another example of things that were constantly in our face while we were trying to focus on coordinating and implementing a major backfire plan."

In the end, the backfire did exactly what Greg had hoped. The fire was prevented from burning $22 million worth of structures along the west end of Gunflint and Magnetic Lakes and on the south side of Gunflint Lake. As the fire came into the burned-out area, it hesitated. Finding no fuel to burn, it turned east. The fire had been turned, but it had not been stopped. While they had hoped it might burn itself out, the fire and the weather conditions had other ideas.

26 HAM LAKE FIRE, DAYS 6–11

Over the next couple of days the Type I team continued flowing into the area. Many of the professionals and volunteers who had been working on the fire since its beginning would continue working on it, though now they would be subsumed into the leadership structure of the Type I team. While the Type II team had three divisions—A (Roach), B (Grindy), and Structure Protection (Lynch)—the Type I team added seven additional divisions to the fray: C, D, E, X, Y, W, and Z. The air attack resources were also expanded with many more aircraft of all types. And since the fire was now in Canada, Canadian firefighting resources were engaging.

On Thursday, May 10, the weather conditions that IMET Byron Paulson had forecasted turned out worse than expected. Incoming cold fronts often bring in rain. This front came in and only brought a shift of strong wind. The red flag warnings that had been predicted were realized. The fire was poised on the north side of Gunflint Lake. As the winds shifted out of the north and gusted up to 30 miles per hour, they pushed the fire east around Gunflint Lake and then almost due south for nearly eleven miles. The fire breached the Gunflint Trail in a nearly two-mile stretch, through Little Mayhew Lake (just north of the Trail), through the old Gunflint Trail (just south of the present Trail), all the way south to almost a mile below

Rush Lake, well inside the BWCAW. Here, too, it burned extensive forest but no more structures. The fire had made its most dramatic run, consuming nearly 30,000 acres in one day.

The fire raged for five more days, during which the wind finally began to diminish, enabling the Type I team to begin putting the fire out. It was a long and exhaustive process.

In the end, almost all of the 144 structures that burned during the Ham Lake fire were consumed on that fateful Sunday and Monday, when the blaze burned through the more populated end of the Trail.

"We used to have eleven homes on our road," commented Corrine Sierakowski in a *Pioneer Press* article published on May 8. "There are three left. There are places that are a total wipe-out, and next door, green trees and grass."

"It's really shocking," said Deb Mark, owner of Seagull Canoe Outfitters, who was quoted in the same article. "It's a numbing experience to go through green one day and black the next."

When asked if he had lost something in the fire, Earl Cypher snapped, "I didn't lose something. I lost everything."

With the notable exception of a handful of rustic cabins on the Canadian side of Gunflint Lake, which were burned on May 11, no other structures were lost. After May 11, the fire burned for another four days, consuming an additional nearly 20,000 acres of forest.

During every day of the fire, Voyageur's Sue Prom faithfully updated the public through her blog. Some of these entries are very long and detailed. On May 12, three days after the big burn, she posted a lengthy entry of more than a thousand words titled "Ham Lake Luck." She referenced the start of the fire and conveyed a sentiment held by many along the Gunflint Trail:

> The acreage burned on the Ham Lake Fire is over 55,000 acres. Half of the Gunflint Trail has been evacuated and the fire is only 5 percent contained. Can you imagine being the person responsible for this fire? The loss of property, the displacement of the

evacuees, the disruption of all of these lives, the risk of the lives of the firefighters? I feel for that person because the weight on their shoulders must be almost too heavy to carry.

It could have happened to anyone. Have you ever left a campfire burning unattended overnight? I know I have. Have you left a campsite for a quick paddle out to try to catch some fish with the fire pit still smoking? I know I have. Have you packed up all of your belongings without being positively sure your campfire was cool to the touch? I know I have.

The person responsible for this fire didn't try to start it. It just happened. The conditions of the forest were primed for fire, and maybe that person did everything they were supposed to do. Maybe the fire burned deep down into the duff, or an errant spark blew into some dry brush in the woods waiting to ignite when the conditions became volatile. Whatever happened, it happened.

We're all great Monday Night Quarterbacks. "There should have been a fire ban in place." Sure, I agree after a fire has consumed 55,000 acres, but if you would have asked me a week ago if I wanted a fire ban I doubt I would have agreed.

The ice just went off of the lakes. The late season snow had kept the Gunflint Trail in a condition better than the rest of the State of Minnesota. The water is cold in May and the air temperatures are normally cold too. Not many people visit the BWCAW in May and everyone who loves camping loves to have a campfire.

Eventually, on May 15, the fire would be largely extinguished, although parts of the forest would continue to smolder for several weeks. By the time the fire was finally subdued, it had consumed an estimated 75,551 acres: the largest forest fire in Minnesota history in almost a century. The more than one thousand firefighters, volunteers, and others who had given so much of their time and sweat to end it were, like the fire, finally spent.

AFTERMATH

27 THE INVESTIGATION

After USFS law enforcement officer (LEO) Barry Huber interviewed Steve Posniak at the Tuscarora Lodge, the night the fire started, he felt certain he had identified the person who had ignited the Ham Lake fire. In fact he was so certain he now had a suspect, he decided to recruit additional investigation support. At some point very soon after first meeting Steve Posniak, LEO Huber contacted USFS Special Agent (SA) Keith McAuliffe in the Duluth office. SA McAuliffe was a lead investigator for Region 9 of the USFS, which included the Gunflint Ranger District.

Coincidentally, LEO Huber had recently been through training for arson investigation with DNR conservation officer (CO) and arson investigator Michael Scott and DNR fire investigator Curt Cogan. While CO Scott and Fire Investigator Cogan worked for the DNR, they participated in arson investigation training for several different agencies. Huber had attended this training to obtain his classroom work as an arson investigator. After attending a weeklong course, attendees still need to perform work in the field in order to obtain their fire investigation certification. Huber may not have had his certification by this time. Regardless, he or SA McAuliffe or both decided they would like the additional law enforcement and fire investigation support from two DNR recognized experts with substantial experience.

Huber contacted both McAuliffe and Scott the night of May 5. The initial contact still needed to work through formal channels. Once Scott was notified, he knew he and Cogan would need to be officially called up by the MIFC. One or both of them contacted the MIFC to let them know.

Cogan, a DNR fire investigator with whom Scott had worked in the past, had been working for the DNR for almost twenty-nine years, most of it in fire investigation. In fact, one of Cogan's jobs over the years was as the forestry enforcement coordinator for the DNR's Division of Forestry. In general, McAuliffe and Scott worked on the overall criminal investigation of the fire, while Huber, Cogan, and LEO David Spain worked the actual fire investigation, though there was overlap between work done by all four investigators.

Because the MIFC had called up Scott and Cogan, they were assigned to John Stegmeir's Type II team, though largely on a dotted-line basis. Huber and McAuliffe were both assigned to the case because the fire had started on federal land and they were the federal investigators. But the four of them worked together to tackle many of the details surrounding the investigation.

On Sunday, May 6, Huber began his day by searching for Steve Posniak's rented SUV in Grand Marais. Fortunately, Grand Marais is not a large town, so it did not take Huber long to find Posniak's SUV parked at the Grand Marais Best Western. Huber inquired at the front desk and found out Posniak was staying in Room 108. Huber called his room, and when Posniak answered, they made plans to meet in his room at 9:45 a.m. Huber needed to take his photograph, but of course he had other important questions to ask him.

When Huber arrived at Posniak's room, one of the first things he noticed was Posniak's overall appearance. "He was cleaned up, but he didn't look any better," Huber said, referencing how Posniak appeared the night before when he first met him. "He clearly did not get a wink of sleep all night. He looked distraught."

Perhaps because he wanted Posniak to understand the gravity of the situation and also to make certain the investigation continued by the book, one of the first things LEO Huber did was read Posniak his Miranda rights.

"He nodded that he understood," Huber said, when describing the incident.

"So here's your question," Huber continued to Posniak. "Tell me what happened, and tell me the truth. Don't try to deny or lie your way out of this."

"He took a deep breath, heavy sigh, and said, 'Yes, I never meant for this to happen. It was an accident.'"

"OK," Huber answered. "I appreciate your honesty. That's all I need to hear. There will be some officers or agents in touch with you."

Huber found out Posniak would be staying in Minneapolis for a few more days. Posniak gave Huber his Minneapolis address, and then, presumably, shortly thereafter Huber departed.

Officer Huber arranged to meet both CO Scott and SA McAuliffe on Monday, May 7, in Duluth. Meanwhile, he needed to gather as much information as he could about where and how the fire started. He began by donning his Nomex gear and heading out to Posniak's campsite on Ham Lake. He gathered evidence and took photographs and noticed several pieces of tissue blowing around.

On Monday morning Huber drove to Duluth.

When interviewed, Huber described McAuliffe as a "suit and tie guy." He was one of a handful of USFS officers who handled only felony-level investigations. Because by Monday, May 7, there had already been structure fires and substantial expenditures fighting the fire, Posniak, given what they already knew, would very likely be charged.

"To see [McAuliffe] work," Huber said, "is amazing." Once Huber explained the entire situation to him and shared the photographs and additional evidence with him, including Posniak's admission of guilt, Huber witnessed McAuliffe's acceleration into investigatory high gear. "He was talking on the phone, working on his computers, barking orders, everybody was scrambling at his commands. . . . It was awesome. We had a lot of stuff done by that afternoon."

While Huber returned to Grand Marais to continue his work with DNR fire investigator Cogan, Scott and McAuliffe visited Posniak at Dianne Runnels's house in Minneapolis. Runnels, a widow, had known Posniak a long time. Her husband had taught with Posniak when they were both at Carthage College. In the past, whenever Posniak visited the BWCAW for his annual trip, he would stay with his old friend Runnels either before or after his trip north. On this trip, he was staying with Runnels's widow after his trip. On Tuesday, May 8, Runnels was not home when Scott and McAuliffe paid Posniak a visit.

"We generally do in-person interviews," explained Scott, especially in instances like this one in which there were maps and related material they wanted to review with Posniak. The idea was to go through the whole process of how the fire started and to take Posniak back to the incident and what took place.

Upon entering the home, one of the first things Scott noticed were the newspapers scattered around the room. The television was on, turned to news. Clearly, Posniak was following the details of the fire. "He was very emotional about what was going on with the fire," Scott commented. "It had taken off and burned a lot of stuff. A lot of land and many structures and so clearly there was a lot of monetary damage involved, not to mention suppression costs to get the fire under control."

As Scott and McAuliffe continued their interview, they uncovered several important aspects to the incident. First, it was very dry, warm, and windy, and the elements were perfect for a fire. Posniak knew about going into the BWCAW and the regulations regarding fires. While the USFS had not yet implemented a fire ban, Posniak knew burning his paper trash was against regulations.

Over the course of their conversation, Posniak admitted burning prohibited materials and that he had started the fire. He also explained that burning his paper trash on his last day in the woods had been his tradition.

"His statement is," Scott continued, "he started the fire in the grate because he's done this every time he's been up there. He turned around and went back to his tent . . . he was doing something . . . and when he turned back, he said the fire was going. There were embers or parts and pieces from the fire itself coming out that were still smoldering, and when it lands it creates a hot spot. The smoldering started the fire. He admitted starting the fire, though not intentionally . . . but through his own negligence."

They also referenced Posniak's permit, on which Andy Ahrendt had written, "Be careful with campfires."

Finally, Posniak expressed his anger to Scott and McAuliffe. Clearly he was emotional but also angry at comments made by Minnesota's Governor Pawlenty. By Tuesday, May 8, the fire was already a huge event, and Governor Pawlenty was visiting the site. Posniak purportedly heard Pawlenty say something like "whoever started this fire is extremely stupid." In fact, the *Hastings Star Gazette* had run an article on May 8 with the following line: "The governor criticized 'the knucklehead who started this' rapidly growing forest fire by leaving an unauthorized campfire."

Obviously Steve Posniak took that comment personally and was upset by it.

In Grand Marais, Huber, DNR fire investigator Cogan, and LEO Spain were busy. Huber began documenting his May 5 interviews with the USFS campers and personnel and others.

On May 7 or 8, Cogan and Spain paddled and portaged into Posniak's campsite on Ham Lake. Even though Huber had already visited the site, taken photographs, and made other observations, Cogan was the fire investigation expert, having conducted numerous fire investigations all over the state.

"There is a standard format you go through on wildfire investigations," explained Cogan. "First, you don't want to jump to any conclusions. Obviously, the campfire was suspicious from the moment we first saw it. That's where most fires get started in the BWCA. You take a broad walk around the site several times. You're looking at burn pattern indicators on the trees, rocks, even tiny grass stems. That process can take you back to the point of the fire's origin. It takes quite a long while to do that because the fire swirls around and winds shift, and you have to stand back and decide what and where to look. On the Ham Lake fire, it didn't start at the campfire grate. It started some distance away. The wind was swirling around there. There were some different factors involved as to how I think it got back to where it did."

On May 8, Huber took USFS Air Resource Management Program employee Trent Wickman's written statement about what happened on Ham Lake.

On May 9, USFS Beaver pilot Dean Lee flew Spain and Cogan into the BWCAW south of Tuscarora Lodge. As part of Cogan's work on the fire investigation, they were taking some aerial photos of Posniak's campsite and the surrounding area. During this flight they located the Ohio canoeists, Eric Mercer and Ted Nichpor, encamped on an island on Tuscarora Lake. Since they were eyewitnesses and reportedly the last ones to see Posniak before he started the fire, Cogan and Spain thought it would be helpful to interview them.

Lee landed the aircraft and taxied to the island where Nichpor and Mercer were camped. LEO Spain's written account described what happened next.

"As we exited the aircraft, either Nichpor or Mercer jokingly asked if we needed to use the privy. They also immediately went into the fact that they had observed a guy with a large bonfire back on Ham Lake a few days ago.

"We separated them, and I asked Nichpor to explain what he had observed."

Cogan interviewed Mercer.

On May 9, SA McAuliffe interviewed three of the USFS Air Resource Management Program campers on Ham Lake, at least one of whom had spoken with Posniak. These were phone interviews, since these people had returned to their homes in Washington, D.C., Wyoming, and elsewhere.

On May 10, Mercer and Nichpor were again interviewed, this time by Scott and Huber. The canoeists were on their way out of town back to Ohio. They were interviewed at the Ranger District Office in Grand Marais. Mercer and Nichpor also submitted their written accounts of what happened.

On May 10, McAuliffe also interviewed another member of the USFS Air Resource Management Program group.

Through the rest of May and into the fall, numerous other interviews were conducted of anyone who knew anything about how the fire started or about Steve Posniak. Investigators pursued leads and witnesses, and either made notes of their interviews or obtained signed, written statements from numerous people involved with the start of the fire, or with Steve Posniak, and with others. They also gathered a raft of related material, including Posniak's initial e-mail orders for supplies to Tuscarora, his meal plan, and the nature journal entries of Eric Mercer, one of the Ohio canoeists.

On March 1, 2008, nearly ten months after the end of the fire, SA McAuliffe and CO Scott visited Dianne Runnels, the person with whom Steve Posniak stayed in Minneapolis before he returned to Washington, D.C. Runnels shared several interesting perspectives about Posniak with the investigators, who were surprised to find out she knew nothing about their visit on May 8, the Tuesday after the fire started. She also knew nothing about Posniak's involvement with the Ham Lake fire. Apparently Posniak never mentioned either their May 8 visit or his role in starting the conflagration.

Considered as a whole, there is much redundancy in the witness interviews about how the fire started and progressed. For example, Mercer and Nichpor were interviewed twice, once on the island in Tuscarora Lake and once in Grand Marais, and both provided written statements, all pretty much reiterating their recollection of events.

The seven USFS Air Resource Management Program colleagues and Kim Wickman (Trent's wife) each had their own perspective on their trip into and camping on Ham Lake, including slight variations in time lines. But more or less they all report the same event, using much the same descriptions and times.

Perhaps the most interesting aspect of these reports is what can be read between the lines, and what it tells us about Steve Posniak's perspective about what he did. He loved the BWCAW and had been visiting the area at the same time each year for more than twenty-five years. At times he appears to have felt embarrassed, angry, shocked, foolish, frightened, or worse about what he did. So much so, he must have spent the afternoon of May 5 in the backwaters of the area around his campsite, conjuring an implausible story in the hopes of avoiding identification and responsibility for what he had done.

Clearly he was agitated, and clearly he felt shame. On the morning of Sunday, May 6, he admitted his role in starting the fire—an accident—to LEO Huber. Later that day, around two o'clock, he visited his old friend Kerry Leeds, the owner of Tuscarora Lodge and Canoe Outfitters from 1974 to 2004. Posniak had first met Leeds in 1979, and presumably had visited almost every year since, even after Leeds was retired.

SA McAuliffe and USFS Special Agent Mary King interviewed Leeds on May 17, 2007. In part, McAuliffe's report noted that "during this visit, Leeds attempted to discuss Posniak's camping trip as the two of them had done many times in the past, but Posniak did not want to talk about it, changing the subject instead. Posniak did reiterate what he had said the night before [on a phone call to Leeds and to others], that he had been staying at Cross Bay Lake for three days."

Even after admitting where he was camped and his role in starting the fire, knowing the officials now more or less had the whole story, he continued his prevarication to others. Again, presumably because he felt shame, regret, and embarrassment.

• • •

Once all the hard investigatory work was done, CO Scott and SA McAuliffe worked with Assistant U.S. Attorney William J. Otteson at the U.S. Attorney's Office in St. Paul to determine whether any charges should be filed in the case and, if so, what those charges would be.

No one could have foretold it was going to be nearly eighteen months before an indictment was finally handed down. In fairness to the U.S. Attorney's Office, not all cases are considered equal. While igniting a fire that destroyed substantial personal property and forest is serious, lives were not lost, and there were no substantial injuries. The U.S. Attorney's Office deals with homicides, narcotics, sex crimes, and more. Presumably these more serious crimes consume attention and resources ahead of what many might consider less serious crimes.

But work on the case eventually did get done, and on October 20, 2008, the first official step was taken.

§28 THE INDICTMENT

There is no doubt that Steve Posniak was responsible for starting the Ham Lake fire. There is also no doubt he violated regulations by burning paper in the Superior National Forest. Since the fire began on USFS land, any timber that was destroyed by the fire was federal property. If he was going to be charged for destroying federal property, by definition it would be a federal crime. Federal crimes must be adjudicated by the U.S. Attorney's Office, and the process to charge, convict, and sentence—at least in abstract—is relatively simple.

1. First, someone must be suspected of committing a crime.

2. A law enforcement officer investigates and works with an assistant U.S. attorney to determine if he or she thinks they should prosecute.

3. Before an assistant U.S. attorney can begin legal proceedings, a grand jury must be convinced the evidence is sufficient to indict.

4. Grand juries hear the evidence and decide if a person (or entity, in the case of a corporation, etc.) should be charged with a crime or crimes.

5. If twelve of sixteen grand jurors vote to indict, the person is charged, and legal proceedings begin.

6. The conclusion of legal proceedings is a guilty or not-guilty verdict.

7. If guilty, a judge determines the appropriate sentence, following federal sentencing guidelines.

Simple enough, but when applied to a specific instance and person —for example, Stephen George Posniak and the Ham Lake fire— the process becomes much more time consuming, nuanced, and complex.

To reiterate, USFS law enforcement officers Barry Huber and David Spain conducted their early investigations the day the fire began (May 5, 2007) and over the days following the start of the fire. Huber also enlisted the assistance of SA McAuliffe, CO Scott, and fire investigator Cogan. After just a few days of the investigation, the team had determined where the fire started, how it started, and who was responsible.

It is fair to say Stephen Posniak's role in the Ham Lake fire was known within the first twenty-four hours of his striking the match that started the blaze. What wasn't known was how long the fire would burn, how much wilderness and how many structures would be destroyed, and the total cost of fighting the fire (and of its destruction).

The extent and expense of the damage from the fire were clearly on Posniak's mind well before the fire was finally put out. On the Tuesday McAuliffe and Scott visited Posniak in the Twin Cities, it was easy to see Posniak's concern. "He was very emotional about what was going on with the fire," Scott said. "When we got to his place, Mr. Posniak had newspapers scattered around, he had the TVs on. He was watching what was going on with the fire. He was just fixated with it. . . . It had taken off and burned a lot of stuff. A lot of land and almost burned down Tuscarora . . . a lot of monetary damage, not to mention suppression costs to get the fire under control. Not only here but in Canada. I know there was some stuff he was very concerned about . . . what the charge was going to be, what the fine amounts were going to be. And a lot of questions we deferred to the prosecutor."

The investigators were, in essence, only doing their jobs: they were trying to determine exactly what happened when and by whom. Questions about prosecution, potential damages, and more were the domain of the lawyers and the legal system, not the professionals who were trying to understand and document what happened.

Over the next year, McAuliffe worked with Assistant U.S. Attorney William J. Otteson, who was ultimately responsible for determining whether the United States should move forward with the Posniak prosecution. Once an assistant U.S. attorney believes there is enough evidence to try a case, he or she must first convince twelve of at least sixteen grand jurors to indict.

According to a pamphlet describing the process (U.S. Attorney's Office, District of Minnesota, *Federal Criminal Prosecution*), "Assistant U.S. Attorneys appear before the grand jury to establish probable cause that a particular person committed a federal felony. They do this by calling witnesses and presenting evidence. . . . Defense attorneys are not allowed to appear before the grand jury; the accused does not need to testify before the grand jury; and the work of the grand jury is to be kept secret."

Sometime during the year following the Ham Lake fire, Assistant U.S. Attorney Otteson presented enough evidence to convince twelve jurors that Stephen Posniak should be charged with a criminal felony. As previously noted, because the grand jury's work is secret, we do not know when they met, who was on the jury, what was said, who said it, or what other evidence was presented. Because of this kind of secrecy and lack of oversight, many have criticized grand juries and their work.

In January 1985, former New York State Chief Judge Sol Wachtler was reported to have said, "By and large [district attorneys could get a grand jury to] indict a ham sandwich." He went on to say, "[Grand juries] operate more often as the prosecutor's pawn than the citizen's shield." Judge Wachtler's "ham sandwich" comment struck a nerve and went viral, at least within the legal community; it is often referenced when discussing how easy it is to convince a grand jury to indict.

Between the investigatory work performed to identify Posniak's role in the Ham Lake fire, and the U.S. grand jury's indictment, which occurred on October 20, 2008, well over one year later, there may have been additional research conducted. But with regard to Posniak's involvement with the Ham Lake fire, everything with which he was finally charged was more or less known the day after he struck the match.

On October 20, 2008, a U.S. grand jury from the District of Minnesota recommended charging Posniak with a three-count indictment. In part, the indictment reads:

THE UNITED STATES GRAND JURY CHARGES THAT:

COUNT 1
(Setting Timber Afire)

On or about May 5, 2007, in the State and
District of Minnesota, the defendant,
STEPHEN GEORGE POSNIAK,

did willfully and without authority set on fire timber,
underbrush, grass, and other inflammable material upon
lands owned by the United States within the Superior
National Forest, namely: the defendant burned paper
trash and other items that ignited a fire in the Superior
National Forest that burned approximately 75,000 acres
in the United States and Canada, and resulted in fire
suppression costs of approximately $11 million, in viola-
tion of Title 18, United States Code, Section 1855.

COUNT 2
(Leaving Fire Unattended and Unextinguished)

On or about May 5, 2007, in the State and
District of Minnesota, the defendant,
STEPHEN GEORGE POSNIAK,

having kindled and caused to be kindled a fire in and
near forest, timber, and other inflammable material upon
lands owned and controlled by the United States within
the Superior National Forest, did leave said fire without
totally extinguishing the same, and did permit and suffer
said fire to burn and spread beyond his control and to
burn unattended, in violation of Title 18, United States
Code, Section 1856.

COUNT 3
(Giving False Information to a Forest Officer)

On or about May 5, 2007, in the State and
District of Minnesota, the defendant,
STEPHEN GEORGE POSNIAK,

did knowingly give false, fictitious, and fraudulent
information to U.S. Forest Service Officers Barry Huber
and David Spain, who were engaged in the performance of
their official duties, namely: the defendant stated that
he camped overnight on Cross Bay Lake, not Ham Lake, on
the evening of May 4, 2007, and that he encountered a
fire already burning out of control at a Ham Lake campsite
on the morning of May 5, 2007 while paddling back through
Ham Lake to Tuscarora Lodge, when in truth and in fact,
as the defendant knew, he camped overnight on Ham Lake on
the evening of May 4, 2007, and he started a paper trash
fire at his Ham Lake campsite on the morning of May 5,
2007 that spread to nearby timber, underbrush, grass, and
other inflammable material; all in violation of Title 16,
United States Code, Section 551, and Title 36, Code of
Federal Regulations, Section 261.3(b).

When there are multiple counts in an indictment, each count has to be considered and adjudicated separately. If the person (or entity) being indicted is found guilty on any of the counts, he (or she or it) must face some kind of punishment for each count. In the case of Posniak's indictment, the first count was a felony, and the second two counts misdemeanors. If Posniak was found guilty for the felony count—"Setting Timber Afire"—the maximum sentence would be six years in prison, and he could be forced to pay some kind of remuneration. He could also be subject to civil law suits seeking remuneration for timber and structure destruction.

Because of his age (sixty-four) and background, never having committed a crime, prison time might have been waived or reduced. If he had had to serve, it would have been difficult. But the prison time might have paled in comparison to the prospect of remuneration, which clearly worried Posniak. With property damages estimated at more than $100 million and firefighting costs an estimated $11 million, he may have believed his retirement savings would be wiped out, and he and his wife would be left destitute. However, that is not what actually happens in cases like these.

If Posniak were to be found guilty of committing a felony, he would be liable for damages. And while courts are keen to punish, they are loathe to destroy. When discussing the possible judgment against Posniak, LEO Huber explained that even if Posniak had been found guilty of committing a felony, the court might have levied a fine but not wiped him out. "The reality of these big fires," explained Huber, "is they're stuck with a huge bill, and they reveal their assets, and they look at what they can reasonably afford to pay, and a judge will look at his situation and determine what he'll pay." In the end, he would have been left with enough income to survive. But it is unclear if Posniak was aware of these kinds of judicial determinations.

From the initial "Grand Jury Indictment" (document 1), filed October 20, 2008, to the final "Order Dismissing Indictment" (document 38), filed January 12, 2009, there was a raft of motions filed in the legal proceeding. They started with a "Summons Issued" for Posniak to appear at the U.S. Courthouse in Duluth, Minnesota, on November 6, 2008, to hear the three counts of the indictment against him. Attending the initial hearing was the plaintiff, Assistant U.S. Attorney William J. Otteson; the defendant, Stephen Posniak; and the defendant's attorney, Mark D. Larsen. The defendant was made aware of the charges against him and entered a "not guilty" plea for all three counts of the indictment. The initial hearing also identified the dates for various court proceedings and motions, including the trial start date of January 5, 2009. Finally, Posniak was released on his own recognizance but forced to give up his passport to ensure he would not leave the country.

Understandably, people respond to judicial proceedings with a variety of emotions. If one has never been in a federal courthouse, simply walking into one does not necessarily stir an appreciation for Renaissance Revival architecture, which is the style of the Duluth federal courthouse where Posniak's initial hearing occurred. The Gerald W. Heaney Federal Building and U.S. Courthouse and Customhouse in Duluth was built in 1930 to last, with a polished granite face, granite stairs, and abutments. The courtrooms are similarly built out with granite, high ceilings, and plenty of turn-of-the-century splendor. While they may feel as warm as a medieval castle, they cannot help but impress whoever enters them, particularly someone on the receiving end of a U.S. government criminal indictment.

Most of us are unfamiliar with the legal process and have never had the experience of defending ourselves in a U.S. federal court. Presumably, Posniak's attorney, Mark Larsen, knew the process and was capable of navigating it and ably defending his client. Regardless, if *your* actions are the reason *you* are on trial, the full weight of the U.S. government must feel like the stones piled atop Giles Corey after the Salem witch trials. The mere grandiosity of the building and its courtrooms must seem surreal, daunting, and frightening.

• • •

According to Jane Comings, Posniak's wife, Steve drove himself from his home in Washington, D.C., to Duluth, to be present for the summons. On his return drive, he "was near home, turning onto a cloverleaf, and apparently got tired and must have fallen asleep?" Regardless, he went over a median and had some kind of traffic issue that required he lose his license for a while. He never told his wife about the accident or losing his license, which was typical for Posniak.

On November 7, the "Arraignment Order" was filed in the case. Again, it reiterated the trial date, its location (U.S. Courthouse, Minneapolis), and identified the judge (David S. Doty) who would preside over the case. It also laid out other legal particulars in the case, including the presentation of evidence and that "the Defendant shall particularize the bases for any Motion to Suppress such evidence."

During the late fall and into December, motions in the case went back and forth. Importantly, Posniak's lawyer filed motions to get counts one and three of the indictment dismissed. According to Jane Comings, the use of the initial phrase in count one—"did willfully and without authority set on fire timber"—disturbed Posniak, as well as others familiar with the case. Their concern was with the word *willfully*; it implied Posniak started the fire intentionally, or that he was burning trash with the sole intent of starting a major forest fire.

"The reference to 'willful' is legalese," commented Scott. "He didn't go out to intentionally start a fire, but as a direct result of his negligence, you could say he willfully started the fire. Because he had an intent to start this fire to do things in conditions that were not appropriate with things that he shouldn't have been burning."

With regard to count three, Posniak did lie to Huber in the early evening of May 5, the first day of the fire. He did, however, reverse himself and tell the truth the next day to Huber, and the next Tuesday, May 8, to McAuliffe and Scott.

Judge Raymond Erickson took awhile to make decisions regarding these motions and others. But on December 15, he finally issued his "Report and Recommendation." In it, he denied all the motions to dismiss that had been filed in the case, including the motions to dismiss counts one and three.

The news of Judge Erickson's ruling must have been terribly disappointing to Posniak. His wife mentioned in retrospect that in the weeks preceding the legal proceedings, Posniak had been getting his papers and related materials in order. Perhaps it was in anticipation of a ruling against him?

Regardless, on Tuesday, December 16, 2008—one day after Judge Erickson's ruling—Steve Posniak died from a self-inflicted gunshot wound. In the end, he was the only casualty of the Ham Lake fire.

Several of the officials and others affected by the fire expressed mixed feelings about Posniak, his role in starting the fire, and his tragic end. There were some who thought the law enforcement effort to investigate and prosecute him was overly aggressive or demonstrated governmental overreach. That said, it is incumbent on the law enforcement and investigation community to find out how fires started and, when necessary, who started them. In Minnesota, those who are found responsible for starting a fire are also responsible for covering the damages resulting from their actions, to the extent they or their insurers are able to do so.

Others expressed sorrow or regret over not having reached out to Posniak, to let him know they did not blame him for the fire. In fact, this was the sentiment of several who lost cabins to the blaze.

In the aftermath of Posniak's passing, his attorney, Mark Larsen, was widely quoted in the press. "He was a kind, gentle, 64-year-old former federal employee who found himself on the receiving end of a very serious criminal charge that, in our view, was an exercise in overcharging by the U.S. attorney," Larsen said, in a *Pioneer Press* article published the day after Posniak's death. About the charges against Posniak, Larsen added, "He was charged with a very serious felony offense for conduct that we would have proven, at worst, was an accident, and I don't think that helped his outlook."

Finally, in a conversation nearly ten years after the Ham Lake fire, Jane Comings shared a poignant story about Steve and the aftermath of his death. She mentioned they had friends who, like Steve, visited the BWCAW every year. On one of their trips, after Steve's passing, they carried his ashes with them and left them in the wilderness area. Steve's ashes finally came to rest and mingle with the ashes from the fire he started. It is ironic and fitting that the final act of Steve Posniak was to nourish the forests in a region he had come to love.

⟨ EPILOGUE

In conducting research for *Gunflint Burning*, I spoke with many who were either participants in the Ham Lake fire or were or are involved with fighting fires. Most of the people with whom I corresponded by phone or e-mail or interviewed in person are mentioned in the sources and acknowledgments section of this book. In most instances, the people I interviewed had many interesting stories, and not all of them could be included in this book.

For example, at different times the Gunflint Lodge was the ICP and hosted the governor and others on visits to the fire. It was also the place where many slept and were fed. Feeding nearly one thousand firefighters who were working hard in the field for extended lengths of time and burning lots and lots of calories was a story (and feat) in itself, the logistics of which cannot be overstated. On a normal week the Gunflint Lodge received one truck order per week from Upper Lakes Food, its food supplier from Cloquet. As the need to feed hundreds began to unfold, Bruce Kerfoot called the owner and said, "We're hurtin' up here." He explained that they had a sudden influx of many more mouths to feed. The owner, of course, had been following the fire in the media and was concerned.

"How many firefighters are you feeding?" the owner asked. "What do you need and when do you need it?"

Kerfoot let him know, and within twenty-four hours he received a complimentary semitrailer load of food. At that point, they were overloaded with food and no longer had the electricity to keep it cold. When the Upper Lakes Food owner heard that, he followed up the semi full of food with a refrigerator truck.

"That's the kind of community we live in up here," observed Bruce.

On May 11, when the fire burned south over the Gunflint Trail, many cabins were threatened. GTVFD Fire Chief Dan Baumann joined by other volunteer firefighters and engines were instrumental in saving those cabins. The Lutsen volunteer fire department had a new engine that had the ability to spray structures with foam. One of those cabins in the path of the flames was covered peak to ground with foam and came through the fire unscathed.

Many of the firefighters who arrived with the Type I team were from the West and were unfamiliar with both the densely forested terrain and boats, which in an area rich with lakes and streams became key firefighting tools in the battle to quell the blaze. Jim Wiinanen and others spent countless hours giving these firefighters a crash course in paddling, fishing, and boat and water safety.

There was a propane refilling station on Voyageur Canoe Outfitters' property that was a lifeline for the refueling efforts to keep all the sprinkler pumps running. At one point Michael Valentini was working in the area when he noticed a smoldering fire near the refilling station. There was a clear path to the station, and if the fire broached it, the explosion would make many of the propane tank explosions that had already occurred during the fire seem like incidental fireworks in comparison.

"All of a sudden the hill's on fire above the big propane tank," Michael recalled. "Mike [Prom] was in charge of structure protection, so they'd given me a radio, and I called Mike as soon as I saw the fire."

"Valentini to Prom," Michael said over his radio.

No answer.

He tried again. "Valentini to Prom." More urgent.

Again, no answer.

He tried again, presumably with even more energy. "Valentini to Prom!"

Finally Prom answered. "Prom to Valentini. I'm busy. You're going to have to hang on a minute."

"Valentini to Prom," Michael shot back. "If I hang on any longer, your propane tank and your hill are gonna go up because there's a fire here!"

"That's when he called in a chopper to drop water and then sent a crew over right away," Michael laughed.

These stories were just a few of the many that were not included in the book's narrative but that demonstrate how massive and long the firefight was, and also how altruism was and is alive and well along the Gunflint Trail.

Backpacker magazine never got the article Gus Axelson and Layne Kennedy originally set out to write and shoot, but it did get a first-hand account of the trio's survival through the fire and what fires in the future may mean for the region (Gus Axelson, "Minnesota's Boundary Waters Face Extermination by Climate Change," *Backpacker Magazine*, September 2007). Although the episode was not something the team ever wanted to experience again, it was good for work. In addition to the *Backpacker Magazine* article, Gus wrote about the fire for the *New York Times*, *Minnesota Monthly*, and the *Minnesota Conservation Volunteer* (Gus Axelson, "A Woodland Lesson in Fire's Power to Destroy and Build," *New York Times*; Axelson, "Where There's Smoke," *Minnesota Monthly*; Axelson, "Revived by Fire," *Minnesota Conservation Volunteer*). Most of the articles featured Layne's dramatic photographs. Layne also used his photographs in his own work, notably in a book he coauthored with Greg Breining, *Paddle North: Canoeing the Boundary Waters–Quetico Wilderness* (St. Paul: Minnesota Historical Society Press, 2011).

After Lee Frelich's impromptu Gunflint Lodge press conference, he returned to the Twin Cities. In his own words: "I stayed there a couple hours and then went back to the Twin Cities, and media people sent crews to the office for the next two to three days. We filmed a lot of pieces outside Green Hall, in back of the building in front of some coniferous trees." Lee's work in the BWCAW continues, where he has more than 750 research plots. He continues to lobby for burns, which he believes give the area its best chance of holding on to some semblance of the flora and fauna that currently reside there.

Many of the people who were the first to work on the Ham Lake fire are now retired. John Stegmeir, Kurt Schierenbeck, Barry Huber, Dan Grindy, Jody Leidholm, Curt Cogan, Tim Norman, John Wytanis, and many others have stepped away from full-time employment with the state or federal government, depending on their positions. Fortunately, many of them are still engaged in helping their organizations train the next generation of firefighters.

Before taking on this book, I knew very little about fighting fires, in Minnesota or elsewhere. After speaking with the professionals and volunteers who spent many days of their personal energy and time containing and finally stopping the blaze, I acquired a new appreciation (and gratitude) for these fighters, one and all. It is comforting to know that if the unthinkable happens and a fire burns out of control, we have heroes like these we can call for help in our hour of need. Thank you.

SOURCES AND ACKNOWLEDGMENTS

In some ways works of nonfiction involving lots of research and a lengthy time to write are implicitly village projects. Many people are involved, probably few as important or as constant as one's spouse. My wife, Anna McCourt, was involved with this project from the very beginning. Even before I thought about covering the Ham Lake fire, she was with me only days after the event, hiking up the trail to Magnetic Rock, witnessing the world that lay in ash. She was the first to point out shoots of green piercing through the blackened soil, demonstrable proof the fire may have destroyed but rebirth was already under way. It was Anna who, hearing my numerous lame attempts to conjure an appropriate book title, said, "Why don't you call it *Gunflint Burning*?" You can't make that stuff up. I am lucky and thankful to have a partner who loves the outdoors and has a recurring ability to capture the essence of just about anything in a single word or phrase.

The first person to suggest that the Ham Lake fire would be a good subject for a book was Ann Regan, editor in chief of the Minnesota Historical Society Press. I had published my first two books with Ann and the Press. Ann's insight made those books better than what I initially submitted, so of course I took her advice and began looking into the fire. Her instincts were solid. While I did not end up publishing this book with the Minnesota Historical Society Press, I am indebted to Ann as the first person to recommend this subject.

At the University of Minnesota Press I worked with many people to finish this project. First and foremost was acquisitions editor Erik Anderson. Erik is a special guy; continually optimistic and interested, he's the kind of representative you want going out to meet and greet people and explain why they should work with the University of Minnesota Press. He moved from Chicago to the Grand Marais area when he was in high school, a total shock to his system, but one that solidified his love of the north woods, particularly northern Minnesota. Kristian Tvedten helped prep the project and worked with the illustrations. Heather Skinner did an outstanding job as publicist. Emily Hamilton was indispensable in marketing the book. Victor Mingovits leveraged his design talents to create a superb book cover. Mary Keirstead did an outstanding job as copy editor, while Laura Westlund, managing editor; Daniel Ochsner, production and design manager; Maggie Sattler, web marketing manager; and Sarah Barker, production assistant, were also involved. The University of Minnesota Press was a terrific partner in getting this book to print.

I interviewed dozens of people—some numerous times—about their participation in the Ham Lake fire. At the top of the list are those who had firsthand knowledge of the fire, worked to contain it, and suffered from its effects. Early on Deputy Tim Weitz patiently answered everything from what he was doing when that first call came in to explaining the codes in the call log so I could decipher them and understand what was going on. Retired USFS Fire Management Officer Kurt Schierenbeck shared his insight, knowledge, and experience with me on countless occasions. At sixty-one, he is not only still active in the fire community, through training and similar pursuits, but he is also one of the oldest members of a northern hockey team that plays and raises funds for charity. Retired USFS Fire Management Officer Tim Norman was extremely helpful at numerous times throughout my research. He provided key documents and insight into the fire and how the fight unfolded. He also gave me names and contact information for many others who were on the fight. Not surprisingly, all of them told me, when I finally reached them, that if I was speaking with Tim Norman, I was speaking with one of the most knowledgeable people in the field and on that fight.

After I had a solid first draft of the manuscript, the University of Minnesota Press reached out to Peter Leschak to give it a detailed read. Leschak is not only a professional firefighter but also a talented writer; among his books are *Letters from Side Lake* and *Ghosts of the Fireground*. He more than lived up to the challenge, making numerous and detailed suggestions for how the project could be improved, all of which I tried to incorporate. His careful eye and helpful remarks made this a better book. Thanks.

Other USFS, DNR, BIA, and volunteer firefighters and citizens who lent their time and support to my research efforts included:

- Andy and Sue Ahrendt, the proprietors of Tuscarora Lodge and Canoe Outfitters when the fire started. They outfitted Steve Posniak and were extremely forthcoming about how the fire threatened their property on the first day and thereafter.

- Gus Axelson responded to numerous calls and e-mails while I was trying to get a sense of what, exactly, he and his companions did for three days trying to keep out of the fire's path.

- Bob Baker was a GTVFD firefighter who worked extensively on the fire; he was particularly helpful explaining how evacuations are executed along the Gunflint Trail.

- Dan Baumann, the GTVFD chief during most of the fire, explained his exhausting days during the event and shared many dramatic photographs.

- Ron and Keli Berg told me about the Seagull Seven and described how they put out numerous fires on one very long night and for the next day and night after the fire swept past them.

- Linda Bruss at the MNICS Aviation Dispatch helped me understand how air, engine, and personnel resources are dispatched from multiple state and federal entities.

- George Carlson answered many calls and e-mails about his work as a volunteer firefighter and about the sprinkler systems he created that saved so many structures.

- Curt Cogan explained the nuance of fire investigation.

- Jane Comings, Steve Posniak's wife, answered numerous e-mails and telephone calls and was especially helpful sharing some of the discovery documents compiled in the case against her husband.

- Jesse Derscheid related his conversation moments before smelling the first smoke from the Ham Lake fire (before it was named or known). He was one of the first to witness the fire's speed.

- Steve DuChien, volunteer firefighter with the Grand Marais Volunteer Fire Department, was helpful in telling me how the county volunteer firefighters were employed after the first blistering days of the fire.

- Wayne Erickson explained how flying a Beaver through turbulent skies isn't really as difficult as it sounds.

- Sheriff Mark Falk shared many of his experiences while working on the fire and answered several calls about the law enforcement issues surrounding the fire.

- Lee Frelich shared not only his story but also his insights into the northern boreal forests and how climate change and other factors are changing what we have come to know and love as the BWCAW and the Quetico.

- Dan Grindy responded to numerous calls and e-mails and reviewed my coverage of his work as a division supervisor on one of the first days of the fire.

- Vance Hazelton, the first IC on the fire, answered several e-mails and telephone calls and was an excellent resource for understanding how quickly this fire grew into a significant force to be reckoned with.

- Barry Huber spent a long afternoon explaining the evolution of his participation in the fire investigation.

- Tom Kaffine, USFS firefighter, shared details regarding his team's dicey evacuation of the Kek Man.

- Layne Kennedy shared details about his fateful sojourn into the woods with Lee Frelich and Gus Axelson on the day before the fire started. He shared several compelling and dramatic photographs that brought the fire— particularly during those first days—visually alive. The coincidence of having Lee, Gus, and Layne on-site during a time when they were there to document the effects of climate change and fire on the boreal forest was startling and synchronistic.

- Bruce and Sue Kerfoot had lots of stories about the Rescue Squad and the GTVFD, and how everyone on the Gunflint Trail rose to the occasion and provided help and assistance during the massive, long firefight.

- Nancy Koss, Cook County emergency management director at the time of the fire, was extremely helpful in explaining the Cook County response and how the county had worked with several of the participating organizations over the years to create the County Emergency Operations Plan, which would interface with the Northeast Minnesota Emergency Response Plan.

- Don Kufahl responded to numerous requests and gave me a tour of the GTVFD Gunflint Station, explaining the use of some of the equipment there.

- Jody Leidholm explained how air attack works in Minnesota and how it is one of the most important aspects of fighting a fire.

- Tom Lynch, who was one of the first to answer the call and was on the fire throughout, provided excellent insight into how fires are fought in Minnesota, particularly in an area of the state he calls home.

- Deb Mark (proprietor of Seagull Canoe Outfitters and Lakeside Cabins) discussed how she was certain she would lose her place—and how surprised she was that it was still standing when she returned.

- Bob Monehan shared a lot of personal information with me, especially about how he stayed behind and faced the onslaught of the flames as they raged over and around his sprinkler-protected house.

- Byron Paulson, the on-site incident meteorologist, described the work he did on fires (especially the Ham Lake fire) as some of the most rewarding of his career.

- Greg Peterson, a key player in the early fight with the fire, was an excellent resource explaining how the fire burned out of control on days two through four.

- Mike Prom was one of the first on the fire and was a key player throughout. He and his wife, Sue, provided a great deal of information about the start of the fire and its dangerous progression those first three days.

- Sue Prom was both a volunteer firefighter and a frequent reporter on the fire. Through her informative blog posts (some of which were quoted in this book), she kept everyone abreast of the fire on a daily basis.

- Kris Reichenbach is a USFS information officer. Early on she provided basic materials on the fire and helped me set up USFS employee interviews.

- Tom Roach, USFS engine captain, was engaged in fighting the fire from beginning to end. He answered numerous e-mails and telephone calls, provided many photographs of the fire, and was decidedly helpful in understanding several details about the fire.

- Michael Scott provided investigation insight.

- John Silliman, a naturalist at Gunflint Lodge, explained how he was able to help two firefighters navigate Kings Road and other remote trails in the forest around Ham Lake so they could go in and find the USFS professionals encamped on Ham Lake on the first day of the fire.

- Jan Sivertson traded e-mails with me and explained how badly she felt about Steve Posniak's end and how she did not blame him for the start of the fire or the burning of her cabin.

- John Stegmeir was indispensable in explaining how incident command works in Minnesota (and how it worked on this fire).

- Ron Stoffel was one of the first people I contacted about the fire, and he shared many of his contacts with me, many of whom were featured in this book.

- Barb Thompson, USFS firing boss, explained the complex process of aerial fire ignition. During numerous phone calls and e-mails, she was extremely helpful in detailing key events in the fire.

- Michael and Sally Valentini discussed those crazy days as they tried to save structures and navigate the smoke and flames.

- Jim Wiinanen discussed his work saving the Wilderness Canoe Base and the experience of losing some of those structures. He explained his water safety training efforts.

- John Wytanis explained how he helped manage the migration of the fire from a Type III incident to a Type II incident, stepping in when he had to, even though the Gunflint District was not his.

In retrospect, I was naive about the personal impact of wildfires. I have never lost anything to the kind of natural disaster chronicled in this book. Wearing my unschooled empathy on my sleeve, I began contacting people who were involved with fighting the fire or who were significantly impacted by it. Usually people were interested in sharing their stories. But a large minority either ignored my interview requests, never answered my calls, or, if I finally caught them unawares—phoning them when they didn't expect my call and they answered—politely explained that the impact from that fire was so traumatic they would rather not relive it.

I think others (for example, the lead investigator, the U.S. assistant attorney, and others) simply did not see any value in having their pursuit of Stephen Posniak and what transpired on Ham Lake and its aftermath chronicled in a book.

Gunflint Burning uses both nonfiction and fiction, or quasi nonfiction, to tell the story of Minnesota's Ham Lake fire.

I never met Steve Posniak. He died before I ever thought about writing a book about the fire or his participation in it. In sections of this book, I convey Steve's thoughts and actions, but of course I could not and did not interview him to confirm his thinking and what he did. More important, the way I describe him starting the fire is conjecture. To write it, I interviewed people who knew Steve. I spoke with and e-mailed his widow on several occasions. I have read most, if not all, of what has been written about him. I paddled the route he took to get to his camp, and I spent some time at the site. I feel confident that his personal history, the wilderness locations he visited, the weather, the campsite from which the Ham Lake fire started, and more are all factual. And while I believe the way I described Steve starting the fire is one likely scenario, it could have manifested in an entirely different way. In essence, these parts of the book are quasi nonfiction.

Similarly, other details were filled in to give the story an immediacy it would not otherwise have. Most of the people interviewed could not remember their exact words or conversation, though they could tell me the substance of what they said. In these instances, I often quoted them, conjuring their conversation from what they told me they discussed, even though they could not recall their precise words. For example, in the Prologue, I describe Deputy Tim Weitz's visit to Bob Monehan to evacuate him. Tim did not remember running through the sprinklers. But Bob told me he had run the sprinklers for twenty-four hours straight before the fire, so they had to have been operating when Tim visited. Also, Tim and Bob didn't remember their exact words, but both remembered the substance of what they said. In these instances, I have taken liberties with their conversation, though I tried to convey it with accuracy.

Technically, the preceding examples (and others throughout the book) are sprinkled with imagined snippets of conversation, thoughts, and actions that provide *Gunflint Burning* with a sense of immediacy. And while in all instances I tried to be as true as possible to the facts as they were described and related to me, when necessary I sometimes filled in gaps with guesses (again, quasi nonfictions) rather than verified fact.

I cannot think of any instance in which I conjured a conversation that I did not subsequently share with the person being quoted, seeking their confirmation on its accuracy.

The Internet makes research on the numerous print resources about the Ham Lake fire relatively easy, which is a blessing and a curse. It's a blessing because it does not take long to find copious amounts of relevant information. It is a curse because you then need to read all this material. I'm pretty certain I have read or at least seen nearly everything that has been written about this fire. From large papers to small organizational newsletters, blog posts, and PowerPoint presentations, there is a lot out there on this fire, much of it useful.

To everyone who helped me learn more about fighting fires generally and this fire in particular, many thanks. To those less willing to share their information about what happened when, I hope this book has at least accurately portrayed some of the stories of what happened to the people and the forest impacted by the Ham Lake fire.

INDEX

aerial ignition/firing, 247–48; creating
backfires with, 248–50, 259, 260–66,
267–72. *See also* air attacks; burnouts;
prescribed burns

Ahrendt, Andy, 96, 283; evacuating from
Tuscarora, 84–85, 95; owner of Tuscarora, 13, 14, 15; reports fire, 72–73;
saving Tuscarora, 118, 123–24; statement regarding Posniak, 129–30

Ahrendt, Daniel and Shelby, evacuating
from Tuscarora, 84–85, 95

Ahrendt, Sue, 72, 146; attempting to find
Posniak, 84, 85, 103; evacuating from
Tuscarora, 85–86, 95, 100; fire alert
phone tree, 74, 75, 79; owner of Tuscarora, 13, 14, 15; saving Tuscarora,
118, 123–24, 125; statement regarding
Posniak, 128, 129

air attacks, 96, 184; air tanker bases,
80–81; dropping soapy water, 211–
12; fighting fire from rear, 194–95,
205–9; plane strikes eagle, 100–101;
protecting Voyageur's refueling station, 192–93; saving Seagull Guard
Station, 226; saving Tuscarora,
84–86, 86–88, 134; for structure

protection, 87, 164, 205–6, 225;
supporting burnouts, 174–75; Type I
response, 273. *See also* aerial ignition/
firing; fire retardant, bombarding
fire with; Leidholm, Jody, air attack
coordinated by

Alpine Lake fire, 27, 76, 90, 91, 96,
123, 157

Anderson, Duane, injured in Windigo
Lodge fire, 37, 38, 41

ash: falling, 5, 113, 135, 136–38; pillars
of, 232–36, 240; thickening, 5, 67,
159, 175, 221, 227, 242, 244. *See also*
smoke; smoke plume

Axelson, Gus, articles regarding Ham
Lake fire, 301. *See also Backpacker*
magazine group

backfires, 255–72, 274. *See also*
burnouts

Backpacker magazine group, 20–28, 33;
camping on Three Mile Island, 31–32,
46–47, 53–56; stranded at Seagull
Lake Palisades, 35, 137–39, 151–52,
220–22, 242–45; traversing Seagull
Lake, 76–79, 110–14, 242–45

CARY J. GRIFFITH is the author of the nonfiction books *Lost in the Wild* and *Opening Goliath* (winner of a Minnesota Book Award) and the novels *Savage Minnesota* and *Wolves*, which received a Midwest Book Award.